THE
MARINE REEF
AQUARIUM

PHILIP HUNT

BARRON'S

First published in the United States
and Canada in 2008 by
Barron's Educational Series, Inc.

© Copyright 2008
Interpet Publishing Ltd.
All rights reserved.

Originally published in 2008 by
Interpet Publishing.

All inquiries should be addressed to:
Barron's Educational Series, Inc.
250 Wireless Boulvard
Hauppauge, NY 11788
www.barronseduc.com

Library of Congress Control Number:
2006928519

ISBN-13: 978-0-7641-6023-3
ISBN-10: 0-7641-6023-0

Printed in China
9 8 7 6 5 4 3 2 1

Above: *Cleaner shrimps are among the many beautiful and fascinating creatures that can be kept in the reef aquarium.*

Author

Growing up in South Yorkshire, England, in the late 1960s, Phil Hunt became interested in aquariums when he started junior school and found a tropical community tank there. Having pestered his parents successfully for a tank of his own, he then kept freshwater tropicals for several years. After a break of a few years, while he acquired a degree from the University of Leeds and a doctorate from the University of York (neither of which are related to fish), he started keeping reef aquariums in the early 1990s. Since then he has been a regular contributor to fishkeeping magazines, most notably *Practical Fishkeeping*. He lives in Sussex, England, with his wife, daughter, son, and quite a lot of marine life.

Contents

The fascinating world of the reef aquarium

A thriving reef aquarium represents the ultimate triumph of the aquarist's art. Unique in its beauty, it is a living system of unparalleled interest and no little educational value. To observe such an aquarium on a daily basis provides more insight into the functioning of natural ecosystems than any number of college courses, television documentaries, or dry biology textbooks.

To emphasize the educational value of the reef aquarium is to risk overlooking its other virtues. Can any other aquarium match its beauty? It is possible to make a case for the underwater garden of a well-planted tank or the colorful hurly-burly of a system devoted to African rift lake cichlids. But even these pale into insignificance in comparison with the dazzling colors and extraordinary patterns of marine fish flitting among the bizarre forms and sometimes outlandish hues of corals, clams, and other invertebrates.

Add to this the fascinating behavior of many coral reef inhabitants, from cleaner shrimps grooming fish that hang as if hypnotized to receive their treatment, through tiny hermit crabs meticulously trying out snail shells for size before moving in, to clownfish gamboling among the tentacles of anemones, and it is easy to see why the television in a house with a reef aquarium is seldom watched.

Over the last few years, our knowledge of how to create and maintain reef aquariums has advanced

Above: *Clownfish among the tentacles of a coral: one of the fascinating interactions between reef aquarium inhabitants.*

enormously. In part, this is the result of improved communications, meaning that information that once might never have passed from one country to another is now disseminated widely, thanks to the Internet. To some degree it is also the result of more aquarists than ever before becoming interested in keeping marine invertebrates—which have always been the limiting factor— with two results. First, the more heads are applied to a problem, the greater the number of solutions produced. Second, the aquarium industry has applied some of its best brains to the task, in response to a market of increasing size.

The outcome of all this is that it is simpler than ever before to create an aquarium that most aquarists could only have dreamed about 10 or 20 years ago. When we consider that we can now not only keep corals that are still referred to as impossible by some older textbooks,

but that these species can actually be farmed commercially, the scale of our achievements becomes clear.

If you decide to take the plunge into the world of the reef aquarium, you will be embarking on a journey that is easier than it has ever been. However, you should not set out on this path too lightly; it will require inputs of time, effort, and money, and you are almost certain to make frustrating mistakes, although this book aims to reduce the chances of these being too serious.

Keeping a reef aquarium also entails special responsibilities, in that at least some, and probably most, of the animals you keep will have been captured in the wild. You will have a special duty of care to those creatures and the unique environment from which they came. Before you buy your marine fishes, bear in mind that many of them can live at least twice as long as a cat or dog—the humble common clownfish has been known to live 35 years in the aquarium. Corals have even greater potential longevity; some wild colonies are hundreds of years old. You will also owe a great debt to the collectors who, not without some risk to themselves, made it possible for you to keep your fish and invertebrates, and on whose usually unsung efforts almost the whole of this hobby rests.

Choose your fish and invertebrates wisely, care for them with the respect they deserve and they will reward you with years of pleasure as your tank matures. Today, your chances of success with this ultimate form of the aquarium are higher than they have ever been, and your captive reef should be able to fascinate and delight you for decades.

This book provides a practical guide to establishing and maintaining a reef aquarium in your own home, with the emphasis firmly on the practical. For most of us, our aquarium has to fit in with our increasingly busy daily lives and needs to be able to function without requiring huge amounts of space or endless maintenance. With this in mind, the focus throughout the book will be on creating reef aquariums of a reasonable size that do not need cutting edge technology and that any enthusiast will be able to maintain. The development of "natural" systems, using living substrates—rocks and sand collected from reef areas and full of marine life—to provide biological

Below: *Once fully mature, a thriving reef aquarium can be home to a wide range of beautiful fishes and invertebrates.*

filtration is what makes the relatively simple reef aquarium possible. To achieve the same results without these techniques, a reef tank would require a tremendous amount of technology and a huge commitment to maintenance. In consequence, in this book we will only be looking at so-called "natural" aquarium systems—those based around living rock, sand, and algae.

Although some experience in keeping fish, and particularly marine fish, can be helpful to the reef novice, it is by no means essential, and this book is designed to allow even a complete beginner to create a thriving reef aquarium. As such, this is not a book about pushing the boundaries of the reef aquarium. Instead, we will concentrate on techniques and technology that work, and species of fish and invertebrates that have a proven record of adapting well to aquarium life. You will not find any discussion here of corals with

complicated feeding requirements or fish that grow to unmanageable sizes or have extremely specialized needs. Going beyond this, the fish we describe are those that can generally be depended on to live in harmony with the corals and other invertebrates that are usually kept in the aquarium.

We begin by exploring the fabulous underwater world of the coral reef, our inspiration and the source of all our aquarium inhabitants, and then move on to discuss the key elements that make up a successful aquarium system. Next, we look at some potential inhabitants, reviewing a range of great fishes and invertebrates. Having seen how to make a reef aquarium work, and what to keep in it, we guide you step by step through the creation of a real tank, before examining how to maintain it and solve some of the more common problems, to keep it looking at its best. Finally, we review the conservation aspects of the reef aquarium hobby.

Chapter One

The coral reef

To dive or snorkel on a coral reef, or even to see a film or photograph of one, is to be presented with a complex, apparently chaotic tableau of bizarre shapes and bright colors, beautiful to behold but hard to comprehend. Underlying all this beauty and complexity, however, is something relatively simple but remarkable: a partnership between a primitive animal (a coral) and a form of microscopic algae known as zooxanthellae. Without this partnership, tropical coral reefs as we know them would not exist. Why? Because the seas where reefs grow contain very little plankton—at least before a reef becomes established. Their water is very clean and clear; there are few nutrients to encourage the growth of planktonic algae (phytoplankton) and without this, there is little for planktonic animals (zooplankton) to eat, so their populations remain small, too. For a coral without zooxanthellae, this is not a particularly good situation—such corals feed themselves by eating plankton, so little plankton means few corals.

However, one thing that is abundant in these nutrient-poor tropical seas is sunlight and thanks to the clarity of the water (due to the lack of plankton), it penetrates deeper beneath the surface than in more turbid waters. While animals cannot do much with light, plants and algae use it, through the process of photosynthesis, to make biochemical food for themselves.

The sheer abundance of life on a healthy coral reef is astonishing.

The coral reef

Without much in the way of nutrients in the water, algae in the clear tropical seas cannot really make the best of all the sunshine available. The evolutionary quirk that teamed up corals with zooxanthellae made life a lot easier for both of them. The coral obtains some of the biochemical "food" produced from the zooxanthellae's photosynthesis, and the zooxanthellae gain some protection from being eaten and can use the coral's waste products as a kind of fertilizer. The effectiveness of this teamwork can be seen in the tremendous growth rates of corals on tropical reefs.

Another feature of corals—their ability to extract dissolved calcium and carbonates from the water around them and deposit it as a form of limestone—is just as important for reef formation. It is the limestone "skeletons" of corals that create the underlying reef structure.

The creation of a reef

At the beginning of any reef, there is nothing but a clear, warm sunlit shallow sea with a rocky substrate. Drifting in the plankton, the larval forms of some juvenile corals settle on the rocks and start to grow, laying down their limestone skeletons. As they grow, they begin to provide a home for a variety of fish, shrimp, starfish, and other animals; life becomes more concentrated around the corals, as it does around any kind of object in the otherwise empty sea. As the corals grow larger, the lower parts die off, leaving their skeletons, the living polyps being up in the sunlight. The dead areas provide surfaces for other animals, such as sponges, mollusks, and more corals, to colonize. Algae also grow on these areas, and encourage herbivorous animals to live on the embryonic reef.

The action of waves and storms breaks off branches of the corals, which settle on the seabed and create new colonies, like cuttings from a plant. In this way the reef increases in both size and complexity, with more and more animals and plants becoming resident. Over decades, centuries and millennia, the lower zones gradually become a mass of limestone, left behind as the reef grows upwards, but still available to be colonized by those animals that live at greater depths. Eventually, a mature reef, as we see it today, is created. Where once there was just water and rock, an immensely complex, three-dimensional structure teeming with life exists.

Reef types

Coral reefs are very varied, but they have been classified into three basic types; fringing reefs, barrier reefs, and atolls. The different types of reef are closely related and can develop from each other; over time a fringing reef can become a barrier reef, and under the right circumstances, barrier reefs can form atolls.

Fringing reefs are formed when corals grow close to the shore, generally in a line that corresponds (or once corresponded) to the low water mark on spring tides. The reef grows outward from this line, sloping away into deeper water. As the reef grows it tends to move outward, away from the shoreline. The end result of this process, often aided by slow subsidence of the underlying rocks, rising sea levels or a combination of both, is that a shallow, calm area of water known as a lagoon

Below: *The double life of reef corals is exemplified here: the extended tentacles are used for capturing plankton and the golden brown color of the polyps is due to the presence of zooxanthellae, which capture the energy of sunlight to fuel the coral's growth.*

are much more important to the aquarium keeper, in terms of providing useful information as to how best to care for their inhabitants, than whether an animal comes from a fringing reef, barrier reef, or atoll.

So, let us look at those zones, bearing in mind that not all of them are present on all reefs.

The fore-reef wall

Starting from the open sea, the first feature is the fore-reef wall, rising from the depths. Let us explore this, from top to bottom. The reef crest is the top of the living, growing coral reef, which is exposed to the full force of wave action. The upper parts of the reef crest may be exposed at low tide. The top of the reef crest is sometimes poorly colonized

separates the reef from the shore. The fringing reef has now become a barrier reef. The final type of reef, the atoll, is best thought of as a roughly circular barrier reef enclosing a central lagoon, but out in the open sea rather than alongside a coast. Atolls typically arise where oceanic islands have been submerged due to the land sinking or sea level rising; many mark the tops of extinct volcanoes. New islands sometimes arise on atolls, as sand banks accumulate on the reefs.

These three basic types of reef are not mutually exclusive, but can be thought of as the basic building blocks from which larger and more complex reefs are composed. For example, Australia's Great Barrier Reef, the ultimate in coral

Above: In the clear, shallow, sunlit water of a tropical reef, a reef of soft and stony corals is home to a vast range of fishes and invertebrates.

architecture (at least in scale) is, despite its name, a complex mixture of barrier reefs, fringing reefs, and atolls.

Reef habitats

There is a tendency to regard the reef as a single, homogenous habitat. Nothing could be further from the truth. Within a reef are distinct zones, each of which represents a different environment. Some species can be found in several different parts of the reef, whereas others are confined to one zone. The characteristics of these different zones

OBSERVATIONS FROM THE BEAGLE

Our basic understanding of the structure and formation of coral reefs is still derived, for the most part, from an old but very distinguished source. Although far better known for his work on evolution and natural selection, as published in 1859 in the famous "On the Origin of Species by Means of Natural Selection, or the Preservation of Favored Races in the Struggle of Life" *(usually abbreviated to* "The Origin of Species"*), it was Charles Darwin who first developed our understanding of reefs. Based on observations made on his sea voyages on HMS Beagle, his book* "Coral Reefs" *was first published in 1842.*

by corals because of the battering action of the waves; algae may be the main inhabitants. Those corals that are present are usually massive or sturdily branched types that can take the pounding of the waves.

Moving further below the surface, on the outward edge of the reef, the diversity of corals increases. This area is sometimes known as the reef edge, and is primarily inhabited by sturdy hard corals, as it is still exposed to fierce wave action.

The next part of the reef is known as the upper reef slope. This is a favorable area and holds the greatest diversity of life of anywhere on the reef. This zone is at sufficient depth to avoid the fiercest wave action, has stable conditions due to the influence of the open ocean, but is still brightly illuminated, particularly in its upper reaches. It is here that the reef is at its most spectacular. The reef slope may be almost vertical or may shelve gradually into the open sea, depending on the underlying geology.

Descending the reef slope, we reach the deep reef slope. The number and diversity of corals gradually diminishes in response to decreasing light levels,

with different species of corals being found at different depths. Eventually, either the seabed is reached or the reef slope continues into depths that are beyond the reach of corals.

Above: Part of Australia's Great Barrier Reef, seen from the air, divides the open ocean (on the right) from a large, deep lagoon (left) with some coral knolls. Most of the reef area is reef flat, with smaller lagoons within it.

The reef flat

Back at the surface, heading away from the open sea, behind the reef crest we come to the reef flat. It is made up of rubble derived from the reef (broken dead coral branches, for example), often on a base of dead

Left: The reef flat looks like a beautiful coral garden, but is a harsh environment, battered by waves and often exposed at low tide, where only robust corals can survive.

ZONES OF A TYPICAL CORAL REEF

Beaches behind reefs are usually made of coral sand, and are protected from wave action by the reef and the lagoon.

The lagoon has calm waters that may be quite turbid, often only experiencing strong currents when flushed by incoming tides.

The reef flat is an extreme environment. It may be left high and dry at low tide (dotted line), baking in the sun. When submerged it is often subjected to fierce wave action.

The reef crest, at the top of the outer reef slope, is where the ocean meets the reef. Here, corals need to be able to take a real pounding from the waves, and only compact, sturdy species grow.

Substantial coral heads, often colonized by a range of other invertebrates, frequently develop in the lagoon, creating substantial structures (known as "bommies" in some areas) or even secondary reefs inshore of the main reef.

The reef itself is built up of the limestone skeletons of dead corals and other organisms, accumulating over centuries.

coral rock. The water is usually shallow, even at high tide, and the reef flat may dry out in parts at low tide, with rock pools in some areas. The reef flat is a very inhospitable environment, characterized by high temperatures, periodic drying, occasional exposure to fresh water (rain), and the presence of sediment derived from both the shore and the reef. Even when the reef flat is submerged, the water may have little movement because the reef crest breaks the force of incoming waves. Despite all this, many corals, including many familiar aquarium inhabitants, live on the reef flat, possessing a variety of adaptations to help them survive the extreme conditions. On a fringing reef, the reef flat lies between the reef proper and the shoreline. On a barrier reef, the reef flat can be a long way offshore. Between reef flat and shore

on a barrier reef lies one of the most important features of the reef, particularly from the aquarium keeper's viewpoint—the lagoon.

The lagoon

The lagoon has relatively calm water, because of the reef's ability to break the force of incoming waves. Typically the floor of the lagoon is covered in sand and/or rubble, and beds of seaweed and turtle grass may be present. Isolated corals of various types may be found growing on the lagoon bed. Coral knolls, known as bommies in some parts of the world, may be present. These are miniature reef ecosystems in themselves, and are often home to a great diversity of corals. The lagoon is inhabited by a wide range of fish and invertebrate species, and in particular acts as a kind of nursery for the

juveniles of many reef fish, the shallow waters and seaweeds providing some protection from predators. Lagoons may range in size from a few hundred yards across to many miles.

On the seaward side of the lagoon is the leeside face, or back reef, which is like the fore-reef slope but faces the land rather than the open sea. This is an area of prolific coral growth; the water is calmer than on the outer reef slope, and on the upper part of the leeside face corals may be exposed to the air at low tide.

Conditions in the lagoon

Inside the lagoon, the water conditions may be rather different from those on the reef crest and outer reef slope. As well as being much calmer, because it is sheltered from much of the wave action by the reef crest and reef flat, it may be warmer, as the open ocean has less influence on its temperature. The water in the lagoon may also be less clear, because of a higher level of dissolved nutrients leading to a denser population of phytoplankton. Corals that live in the lagoon are adapted to these conditions and are often quite different (in ways that translate to their aquarium requirements) to corals from other parts of the reef.

Atolls

We have mainly looked at fringing reefs and barrier reefs, but the zonation of atolls is generally simpler than for other reefs. Inside the atoll, the central lagoon is usually quite shallow—up to about 65.5 feet deep. Conditions are generally similar to the lagoon of a barrier reef, with isolated patch reefs and coral knolls present in many cases, and a leeside face colonized by similar species to those on a barrier reef. Outside the main reef of the atoll, secondary fringing or barrier reefs may arise, enclosing an outer lagoon. Beyond these lie the reef faces, equivalent to the fore-reef wall of a fringing or barrier reef,

which slope away steeply, often into very deep water.

The complexity of the reef

Irrespective of the type of reef, the most productive zones (the upper reef slopes) share some important characteristics. The first is complexity, both of physical structure and of biological populations. The branching growth patterns of many hard corals create an environment that

Above: Upper reef slopes are home to a great diversity of corals and are often teeming with fish. A greater number of species is found in these areas than in any other part of the reef.

is characterized by extremely complex geometry, which in turn creates a huge surface for algae and other invertebrates to colonize, and a diverse range of smaller habitats within the reef slope.

For example, a few square yards of reef slope may contain large crevices inhabited by moray eels or groupers, flat areas on which damselfish "farm" turfs of algae as food, highly ramified coral heads into which small fish can retreat as predators approach and innumerable small nooks and crannies that provide

Left: Seagrasses are among the few true plants that live in marine environments. Shallow lagoons, sheltered by coral reefs, often have extensive seagrass beds that are home to populations of juvenile reef fishes.

Right: A dense stand of staghorn (Acropora) is home to a large shoal of Chromis and other damselfishes. The fishes feed on plankton and dart back into the safety of the coral branches when predators approach.

a home for worms, shrimps, and so on. There will be dead coral branches on which sponges or soft corals have grown, these in turn providing a home for a variety of other animals, from fish to microscopic crustaceans. The abundance of life on the reef is such that every square inch is fought over in life-and-death struggles by sessile invertebrates, and many reef fish are intensely territorial about their patch of reef, fiercely defending it against all comers, particularly members of their own or similar species which would compete directly with them for food.

From this it is easy to see that the topographical complexity of the reef favors the development of biological complexity, with a vast range of different species of animals and algae, each occupying its own niche in the complex web of life that makes up the reef community. As we have already seen, this complex ecosystem is itself vital for maintaining the reef as an area in which stony corals can grow, and so maintain the integrity of the reef.

The diversity of the reef

To get an idea of just how complex the reef ecosystem is, simply consider two things: first, the diversity of even that small proportion of coral reef fish species that finds its way into the aquarium trade—their vastly different sizes, shapes, colors, feeding habits, and so on. Such diversity can only exist where there is such a range of ecological niches to drive the evolution of so much variation. Next, consider that in the world's richest reefs, those of Australia and the Philippines, there are well over 2,000 different species of fish. Again, only an environment with such a diverse range of food sources and microhabitats could support such a vast number of different species, and remember, this is just the fish. Add in all the invertebrates (including 800 different coral species), plus a few marine mammals and reptiles, and the algae, and you have some idea of just how complicated the reef ecosystem at its most highly developed can be.

Where to find a coral reef

Coral reefs do not grow just anywhere. If you look at a map of the world with the reef areas marked on it, you will notice a few things. First, the reefs are all in tropical or subtropical areas. This is an indication of the temperature requirements of the corals that build the reefs—an average winter water temperature of at least 68°F is needed. The next thing you will notice is that there are no reefs, even in the tropics, near large river deltas, such as those of the Amazon, Orinoco, or Ganges. This is because reef corals have evolved to be dependent on abundant supplies of sunlight. The presence of significant quantities of suspended solids in the water cuts down light penetration tremendously. Large river deltas wash huge amounts of silt into the sea, creating turbid conditions that are inhospitable to corals. Other sources of sediment (for example, soft underlying bedrock) lead to similarly turbid water and, as a consequence, no reefs.

Looking again at the map of the world, there are also areas, such as the western coasts of Central and South America or of Africa, that fall within the latitudes at which coral reefs might be expected, but there are none.

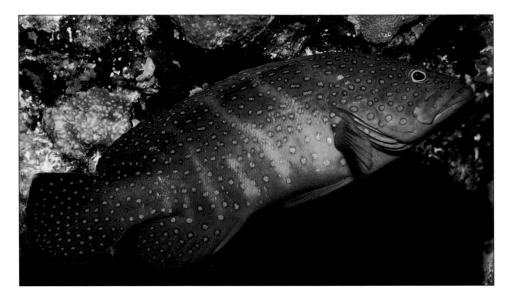

Above: *Unlikely as it seems, the bright colors of this grouper* (Cephalopholis miniata) *may provide camouflage as it lurks among colorful soft corals and sea fans, waiting to ambush smaller fishes.*

The reason can be found in the water temperature. These coasts are swept by cold ocean currents that keep the water temperature below the point at which reef corals can survive.

Reefs in different areas vary considerably in the numbers and types of different creatures that are present. The most biodiverse reefs in the world are found in the Indo-Pacific Ocean, across a huge area including the Philippines, the Great Barrier Reef off Australia, New Guinea, and the islands of the Coral Sea.

The diversity of reefs seems to be dependent on how close they are to other reef areas. Sulawesi, for example, sits at the center of a vast region, stretching from the Andaman Islands to Tonga, with a dense concentration of reefs, and this is where you can find the greatest numbers of different species. The Caribbean, in contrast, is relatively

isolated and despite having one of the world's larger barrier reefs (off Belize), is relatively species-poor. Where reefs are a long way from other reefs, for example in the Red Sea and Hawaii, not only does the number of species tend to decrease, but the number of endemic creatures (those found nowhere else) increases.

The types of creatures found on reefs also varies between different areas. To take just one example, sponges and gorgonians (sea whips and sea fans) in the Caribbean are much more prominent (and colorful) features than on most Pacific reefs.

Aquarium insights

This brief introduction to the wild reef environment can give us an insight into the conditions that we need to provide in the aquarium in order to keep corals successfully, and to some of the issues that might arise. We can see that we need tropical temperatures and that intense light is very important. Water that is low in nutrients is another key factor, as is providing the degree of water motion appropriate to the reef

zone from which the corals of interest come.

Some other requirements are less immediately obvious, but become apparent after a little thought. For example, stony corals and coralline algae create the reef by depositing limestone into their structures, but where do the raw materials for this process come from? Limestone is calcium carbonate, and to create it corals need plentiful supplies of its components, calcium and carbonates. These are abundant in seawater, particularly in areas around coral reefs, and need to be provided in the aquarium. Finally, the intense competition for space and resources on the reef means that we can expect a certain amount of territorial conflict from our aquarium inhabitants, and that we will need to manage this carefully.

COLD WATER CORAL REEFS

Stony corals, and even coral reefs, are not limited to warm seas. Cold water coral reefs also exist, but these are very different to their tropical counterparts. They are found well away from land, at depths of at least 131 feet (and as deep as 3,282 feet), and at temperatures down to 39.2°F. The corals that build these reefs have no zooxanthellae and depend entirely on catching plankton. In sharp contrast to their tropical relatives, they are extremely slow growing (about .04 inches per year). There are some fine examples off the coasts of Scotland and Norway.

CORAL REEFS AROUND THE WORLD

Coral reefs are not evenly distributed around the world, or even around tropical and subtropical areas; specific geographical conditions are required for reefs to form.

In the Americas, most of the coral reefs are found in the Caribbean. The Eastern coasts of South America are swept by cold currents, making them inhospitable for corals. The Galapagos Islands are home to a very limited selection of corals, at the limit of their natural range.

The Red Sea and Persian Gulf are home to the world's most northerly reefs, and the lack of cold currents and few major river deltas along the East African coast enables extensive coral reef growth. Indian Ocean reefs are the source of many aquarium specimens.

Above: *The Red Sea has some of the world's most northerly coral reefs: the winter water temperature is close to the minimum for corals to survive.*

The reefs of Indonesia and the Philippines have the greatest biodiversity in the world's oceans. They are home to thousands of species of fishes and more corals than anywhere else on Earth.

Australia's 1,243 miles Great Barrier Reef (on the East coast) is the world's largest, and there are world-renowned reefs on the West coast, too.

Chapter Two

The aquarium system

To keep coral reef animals in the aquarium, it is essential to provide them with an environment that is similar in a number of important respects to their wild habitat. Water chemistry, temperature, salinity, light intensity, and water movement all need to be carefully controlled and maintained within certain limits in order for the aquarium inhabitants to thrive. The aquarium system is designed to create these conditions and keep them stable.

In this section we look at the key aspects of the aquarium system, from making artificial seawater, keeping the tank at the right temperature, providing the proper degree and type of water movement, and lighting the aquarium so as to provide the intense illumination that corals and other creatures need. We also look at biological filtration and the vital role that live rock can play in managing the waste products of the aquarium inhabitants and keeping the water clean. In addition, we look at supplementary filtration methods, such as the use of protein skimmers and algae filters, as well as considering ways of maintaining other aspects of water quality, such as keeping calcium and carbonate levels high and using chemical filter media to reduce levels of dissolved organic waste products.

Salt and water—the most basic parts of the reef aquarium system

Salt and water

Using natural seawater is not a viable option for the majority of fishkeepers. Artificial seawater, made using commercially available marine salt, is a better choice for most aquarists.

Marine salt

There are many brands of marine salt on the market and in practical terms there is little to choose between them. The salt component of artificial seawater is therefore very easy to obtain but, strange as it may seem, the water often needs more preparation.

Water—and how to clean it up

Mains water supplies, while perfectly adequate for humans, may not be ideal for reef aquarium use, as they often contain levels of nitrates and phosphates (probably originating in fertilizers used in farming) that are too high for many marine creatures. If your tap water falls into this category (which you can establish by testing it), it will need purification before you can use it in the reef aquarium. The three popular ways to do this are by reverse osmosis, deionization, and ion exchange.

THE "SALT MIX" OF NATURAL SEAWATER

Seawater is not simply a solution of sodium chloride; it also contains sulphates, magnesium, calcium, potassium, and many trace elements.

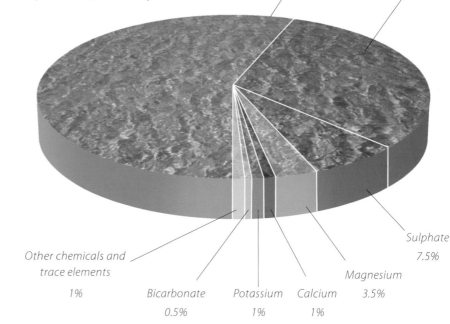

Chlorine 55%

Sodium 30.5%

Sulphate 7.5%

Magnesium 3.5%

Calcium 1%

Potassium 1%

Bicarbonate 0.5%

Other chemicals and trace elements 1%

Reverse osmosis

Reverse osmosis (RO) forces water under pressure through a membrane that allows water molecules to pass through, but holds back larger molecules (nitrate and phosphate ions, for example). Reverse osmosis is very effective, but only a fraction of the water (usually about one-sixth) passes through the membrane and is ready for aquarium use; the rest runs to waste (although it can be used for watering the garden). If you have an abundant, cheap water supply, this is probably the best way to purify water for your reef aquarium. Aquarium RO units are relatively inexpensive and most produce water reasonably quickly in the volumes required for topping up evaporation losses and performing partial water changes. However, with

Left: The fishes, corals, and all the other diverse life on the coral reef depend on the precise chemistry of seawater, and this must be replicated in the aquarium if the creatures are to thrive.

WATER SOURCES

Early marine aquariums used natural seawater, collected from the coast. Indeed, London Zoo's aquarium still does, having water delivered by ship from the Bay of Biscay. More remarkably, from the 1930s until 1959, Chicago's Shedd Aquarium had dedicated rail tankers used for shipping water from Florida.

most RO units, making enough water to fill a large aquarium requires patience. Many aquarium dealers sell RO water.

Deionization

Deionization works on a different principle. It uses resin beads that bind dissolved ions (including nitrate and phosphate ions) in the water passing through the unit, removing them from solution. Deionizers are inexpensive to buy and very effective at purifying water, with virtually no waste, but they do have some disadvantages. Their flow rate is typically very low; this is not much of a problem when making water for topping up evaporation losses, but even preparing water for partial water changes, let alone filling a tank, can be a tedious process when using this method.

Furthermore, the resin beads eventually become saturated with ions, and water then passes through without any purification. Many resins change color when they are exhausted, so at least you can see when this has happened. When the resin is exhausted it must be replaced, and replacement resin is not cheap. The volume of water that a given quantity of resin can process depends on the concentration of ions in the water. If nitrate and phosphate levels are high, the resin will quickly become saturated. Also, hard water (which has high levels of calcium and/or magnesium ions) will saturate deionizing resin more quickly than soft water. Deionizers are a viable method of water purification for aquarists with relatively small systems and consequent small water requirements, and where water is expensive or supplies are limited.

Ion exchange units

Ion exchange units operate in a similar way to deionizers, using resin, except that when they bind ions from water, the resin releases different ions. Typically a nitrate or phosphate ion is replaced

HOW REVERSE OSMOSIS WORKS

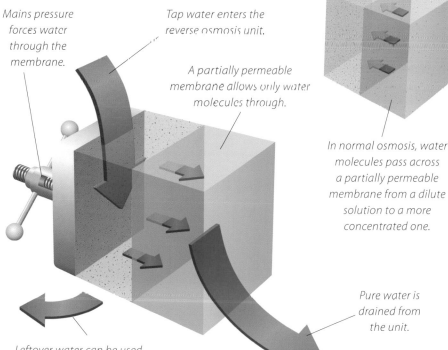

Mains pressure forces water through the membrane.

Tap water enters the reverse osmosis unit.

A partially permeable membrane allows only water molecules through.

In normal osmosis, water molecules pass across a partially permeable membrane from a dilute solution to a more concentrated one.

Pure water is drained from the unit.

Leftover water can be used on the garden.

Above: *A home reverse osmosis system. These can typically produce in the range of 53–132 gallons of very pure water per day.*

with a chloride ion. The water produced using an ion exchange column is not as pure as that produced by RO or deionization. For one thing, it will contain some sodium and/or chloride ions, but these do not matter when the water is being used in a marine system, where a large component of the salt water is sodium chloride. Ion exchange units can produce water relatively quickly with little waste. They are cheap to buy and the resin can be recharged by washing the column with a concentrated sodium chloride (common salt) solution, which reverses the ion exchange process. The water they produce is perfectly adequate for reef aquarium use.

Salinity

In any marine aquarium, it is important to get the salinity, the concentration of salt in the water, correct. The ideal is to match the salinity of the water on a wild reef. This varies between different areas, depending on the geographical circumstances. For example, the Red Sea has water that is much saltier than other areas, because it is essentially landlocked,

Right: A glass hydrometer must be used in a cylinder of tank water and requires careful reading. This design also incorporates a thermometer.

with relatively little exchange of water with the open ocean. In addition, it has a very low rainfall, being surrounded by desert. Most reefs in the open oceans, whether the tropical Atlantic or Pacific, or the Indian Ocean, have a salinity of about 5 ounces of dissolved solids (all the different components together) per gallon (liter) of water, which is equivalent to 35 parts per thousand. This is the usual way of expressing salinity; the corresponding figure for the Red Sea is up to 45 parts per thousand.

In traditional marine fish tanks, salinity was usually kept a little lower than that of natural seawater, at around 30 parts per thousand. Because lower salinities allow higher levels of oxygen to dissolve, they were thought to help prevent fish diseases and were economical with salt. Today, however, it is generally considered to be better to stay close to the salinity of natural seawater, particularly when keeping invertebrates.

Measuring salinity

Although it is possible to measure salinity directly, most aquarists use an indirect method, namely by determining specific gravity (SG) with a hydrometer. Hydrometers are not always highly accurate, but are generally good enough for aquarium use, because it is more important to keep the salinity stable than to match any particular value.

Specific gravity is a ratio of the density of a liquid to the density of pure water. When substances are

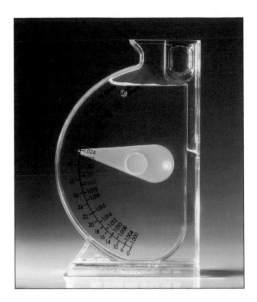

Above: Swing needle hydrometers are easy to read. This model displays both salinity and specific gravity. These hydrometers are claimed to correct for water temperature. Clean them carefully between uses.

dissolved in water, the density of the water increases. The more concentrated a solution, the higher the density and the higher the specific gravity. Density varies with temperature, so specific gravity is properly expressed as being the measurement at a certain temperature.

Traditional hydrometers

Traditional hydrometers are made of glass and shaped like a fishing float. They are floated in the tank water, and the height at which the hydrometer settles indicates the SG. If the water is measured at a different temperature to that at which the hydrometer is calibrated, a conversion table must be used to get an accurate result.

Swing-needle hydrometers

A more recent development is the swing-needle hydrometer. This consists of a slim container that is filled with water from the aquarium. Inside is a pointer that swings up and indicates the specific gravity on a scale marked on the side of the container. Swing-needle hydrometers are claimed not to need correction for temperature, as the plastic that the pointer is made from is designed to change its density with temperature at the same rate as seawater.

Using a refractometer

An alternative method of measuring salinity is to use a refractometer. These devices are not affected by temperature, but do need to be cleaned carefully after use and calibrated periodically by using a standard salt solution. They are more expensive than hydrometers, but also more accurate.

Electronic meters

A final method of measuring salinity is to use an electronic meter. These are still more expensive, but usually compensate for temperature and give very accurate results. Like refractometers, they need to be calibrated regularly.

Right: Dissolving salts in water changes two of its physical properties: electrical conductivity and refractive index. Changes in both can be used to measure salinity. Electronic meters measure conductivity changes, whereas refractometers (right) exploit refractive index changes. Both can provide very accurate readings of salinity, but they need to be calibrated regularly and are more expensive than hydrometers.

Left: This sailfin tang (Zebrasoma desjardinii) comes from the Red Sea, where the salinity is higher than in other coral reef areas. Red Sea fishes adapt well to living at lower salinities than they experience in the wild.

USING A REFRACTOMETER

1 Using a pipette, place a couple of drops of tank water on the viewing window and close the transparent flap.

2 Point the refractometer at a light source and look through the eyepiece to examine the salinity reading.

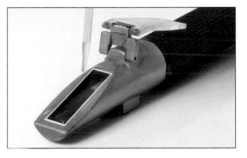

This is the view looking through the eyepiece. The bottom edge of the blue area shows the salinity of the water sample.

The water sample forms a thin film across the surface of the prism. As light passes through the water film, it is refracted and the amount of refraction is proportional to the salinity.

Water movement

Water movement is as much a defining feature of an aquatic environment as temperature or illumination, and getting it right is critical for some reef aquarium inhabitants. Three main types of water movement occur.

• Surge is the back-and-forth motion produced by waves.

• Turbulence is a chaotic movement, with currents swirling in all directions, and is produced when opposing flows meet, for example when the backwash from one wave hits an incoming wave.

• Laminar flow is a steady current in one direction.

Each reef zone has its own characteristic pattern of water movement: powerful surge and turbulence, created by breaking waves, on the reef crest and upper reef slope; laminar flows on deeper reef slopes where waves have less influence; and relatively calm water, with tides providing the only strong currents, in lagoons.

Corals growing in each zone have evolved to live in these patterns of water flow, although some coral species are more adaptable than others and can be found in a wide range of environments. Mushroom anemones and large, fleshy stony corals, such as *Trachyphyllia* and *Cynarina,* that come primarily from lagoons prefer gentle currents, while corals from upper reef slopes, such as *Porites* and *Acropora*, are adapted to vigorous water movement.

This is something to consider when designing the aquarium, as well as when positioning individual corals.

An aquarium intended to house a collection of *Acropora* or similar corals needs strong, surging currents that can sweep unimpeded around the tank. In contrast, an aquarium dedicated to mushroom anemones will require gentler flow and should have quiet areas, in the lee of rocks, for example.

What does it do?

Water movement has several effects on both individual animals and on the aquarium as a whole. For corals, water movement facilitates gas exchange, disposal of waste products, and uptake of calcium and trace elements, as well as delivering planktonic food. It also increases the efficiency of photosynthesis, as the back-and-forth motion produced by surge exposes more coral tissue to light.

For the aquarium as a whole, good water movement ensures that effective biological filtration occurs in live rock and sand beds. The water is also kept well mixed, preventing hot and cold spots and deoxygenated areas; solid

Above: *The reef crest is subjected to very powerful water flows, mainly surge and turbulence generated by waves crashing over it. Corals from this zone are adapted to live in such high-energy conditions.*

wastes are kept in suspension, where they are more likely to be removed by skimmers or caught and eaten by filter feeders; and gas exchange is improved by disturbance of the water surface.

Providing water movement in the reef tank

In the home aquarium, water movement is usually provided by water pumps. If the aquarium has a sump, one source of water flow is the pump returning water to the tank. In aquariums devoted to corals that like gentle currents, this alone might provide enough movement, provided that the minimum flow required for biological filtration is achieved.

For corals that require stronger water movement, more pumps will usually be required, with some kind of controller if surge is to be provided. There are three main types of circulation pumps.

• External circulation pumps are similar to those used to return water from the sump to the tank. They need to be plumbed to the tank via dedicated pipes. This risks leaks, but external pumps have the advantage that because they are not submerged they do not add heat to the aquarium water.

• Powerheads and similar pumps are available in a wide range of sizes. They are usually inexpensive, compact,

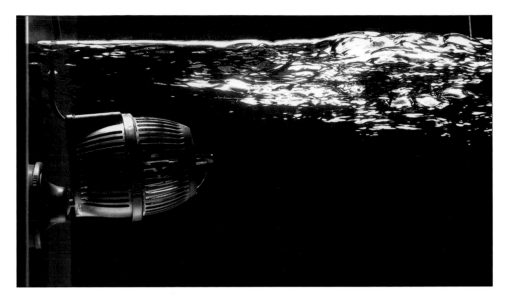

Below: *The water surface in a typical reef aquarium should appear to "boil" as a result of the water movement, with plenty of waves and ripples.*

Above: *Setting up circulation pumps, such as this propeller pump, close to the surface creates fast water flow and disturbs the surface, improving gas exchange in the tank.*

INCREASING GASEOUS EXCHANGE

Disturbing the surface with water currents improves gas exchange by increasing the water surface area and constantly mixing the surface layer with water from deeper in the tank.

● *Carbon dioxide*   ● *Oxygen*

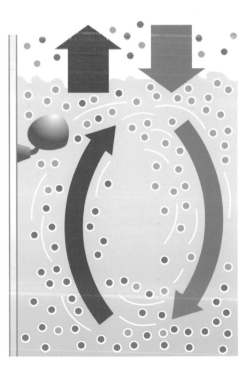

and easy to conceal in the aquarium. These pumps can create strong water movement, but must be run submerged, and so can tend to heat the tank water—particularly the larger, more powerful models. Also, they deliver their water flow as strong, narrow, high-velocity jets, which are unlike the pattern of flow in nature, and which can be damaging to corals placed too close to them. There are two types of these pumps: most powerheads use synchronous motors, whereas some (usually more powerful and expensive) circulation pumps use non-synchronous motors. The difference, in practical terms, relates to how they can be controlled with electronic timers.

• Propeller pumps operate in a similar way to powerheads, but instead of a small impeller to move the water, they have a larger propeller. Very high flow rates can be produced from compact propeller pumps, with low power consumption and relatively little heat output. Propeller pumps produce a broad, powerful stream that provides excellent water movement for corals. Propeller pumps can have either synchronous or non-synchronous motors.

To generate surge, electronic timers (so-called wavemakers) can be used to control circulation pumps. Various models are available that will control one or more pumps; they work by switching the pumps on and off at intervals that can be set by the fishkeeper. Not all pumps can be used with these devices; some can

FITTING A STRAINER

The inlets of circulation pumps should always be fitted with strainers. This is especially important when pumps are being switched on and off by a timer; animals will avoid the suction at the intake of a running pump, but not necessarily when it is switched off. When the pump starts again, fish or invertebrates can be sucked into or through it. Clean strainers periodically to keep pumps working efficiently.

Below: *Strong currents in this large reef aquarium wash the soft corals and gorgonians in the foreground to and fro.*

be damaged by constant turning on and off. Wavemakers for use with non-synchronous pumps vary the flow rate of the pumps, instead of just switching them on and off.

The best water movement

When calculating how much water movement to provide in a reef aquarium, a good rule of thumb is to aim for a minimum total flow rate (which includes the combined flow rates of all the circulation pumps and sump return pumps, if a sump is used) of 10 times the tank volume per hour. This is enough for effective biological filtration in the live rock and sand bed. It is also adequate for those invertebrates that like gentle water movement. For those that need more vigorous currents, total flows of 20–30 times tank volume per hour are more appropriate.

Choose circulation pumps to match the tank size. Although it is possible to use several relatively small pumps to achieve the required total flow rate, it is much better to use one or two larger pumps. For the most effective water movement, pumps (or pump outlets) should be set up close to the water surface. This improves gas exchange in the aquarium by disturbing the water surface. Also, water can move faster at the surface (because the air above offers less resistance than water would), and this allows better circulation than if the pump was deeper under water.

Positioning pumps or outlets at the rear corners, facing out into the center of the tank, works well in most aquarium layouts. Using a single pump, most parts of the tank will receive some flow. If there are two pumps, one in each rear corner, you create a turbulent zone where the flows meet. Using a wavemaker to control either one or two pumps in these positions, you can generate surges across the tank.

FIXING SUBMERSIBLE CIRCULATION PUMPS

Always fix submersible circulation pumps securely into place. If they fall off, the consequences can range from disturbed sand beds to damaged corals. The suction cups supplied with many smaller pumps almost invariably fail, usually quite quickly. Clamps that fix the pumps to the bracing bars or rim of the tank are better, but best of all are magnetic holders. These allow very flexible and extremely secure placement of pumps.

WATER FLOW IN THE AQUARIUM

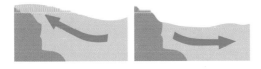

Surge—*water flows in and out*

Turbulence—*random flow*

Laminar flow—*across the face of the reef*

The outlet from the return pump from the sump provides one source of flow.

A dedicated circulation pump, angled so that its flow meets the current from the return pump.

A zone of turbulence is created where the flows meet in the center of the tank.

Heating and cooling the aquarium

Coral reefs are found in tropical and subtropical seas, so their inhabitants are adapted to live in warm water, but at very high temperatures corals begin to suffer. They expel their symbiotic algae (zooxanthellae) and may die. In the wild, unusually high sea temperatures following so-called "El Nino" climatic events have been associated with mass coral bleaching episodes in some areas. In most reef areas, except in very shallow water, temperatures remain fairly stable throughout the day and over the course of the year. Average water temperatures in different reef areas range between 77°F and 86°F, with peaks between 83°F and 94°F.

Heating

These figures provide a guide to optimal reef aquarium temperatures. Traditional fish-only marine aquariums were usually kept at around 75–79°F. This is towards the lower end of the natural range and permits more oxygen to dissolve in the water—a distinct benefit in tanks with poor water circulation and traditional biological filters.

For the reef aquarium, a stable temperature between 81°F and 84°F is ideal. Stability is important; sudden rises can induce coral bleaching and rapid falls can cause outbreaks of fish diseases. In temperate climates, heating is required to keep the temperature steady, particularly in winter. Simple heater-thermostats are ideal for this; they can be placed in the tank or in a sump. The size (power or wattage) of heaters required depends on the size of the aquarium and the ambient temperature. Splitting the total required wattage between two or more units is useful, as it provides some protection against the worst consequences of a

HEATER-THERMOSTAT

Most heater-thermostats are adjusted using a knob on the top of the unit. Modern designs often have a temperature scale marked on them, but water temperature should always be checked using a thermometer.

heater failure. If a single heater unit (with enough power to heat the tank on its own) fails to switch off when it should, it will quickly overheat the tank. If it is one of multiple units of lower power, overheating will take longer, offering a better chance of correcting the problem before it becomes critical. Similarly, if one of multiple units fails to switch on when heating is needed, the tank will still have some heating capacity, so any temperature drop should not be catastrophic.

Aquarium thermometers are available in a range of styles. Floating glass spirit thermometers are not recommended in a reef aquarium, as the strong currents can cause them to detach from the tank glass and they may break if they then hit rocks or corals. You can use inexpensive digital strip thermometers that fix to the outside of the tank, but electronic thermometers, with a temperature

Above: *Submersible heater-thermostats must always be kept completely under water when switched on. Most models have suction cups to fix them to the side or base of the tank or sump, but these usually need replacing frequently. It is better to locate heater-thermostats in the sump (if one is used), rather than in the display aquarium.*

probe inside the tank and a digital display outside, are best.

Keeping cool

Sometimes cooling rather than heating is required. Reef systems (particularly their lights) can generate a great deal of heat, which can overheat the tank, especially during hot weather. There are ways to minimize the chances of this. The tank's location can have a big effect—a cool, well-ventilated room is best. Cooling the lights with fans is effective, as is timing the peak light intensity to coincide with cooler times of day (perhaps splitting the time of peak intensity between morning and evening).

If such measures are not enough, there are two more drastic approaches. Both will have a big impact on electricity bills. The room housing the tank can be cooled using a domestic air-conditioning unit or the tank can be cooled using an aquarium chiller. Chillers can be very effective, but need to be used carefully and may not fit into every domestic situation. Select a chiller according to tank size and the degree of cooling required. It must be supplied with water using a dedicated pump of the correct flow rate. Position the unit with care. Chillers work by removing heat from the aquarium water and transferring it to the external environment. If a chiller is located somewhere that is too warm, too badly ventilated, or too close to the aquarium (an aquarium cabinet is a perfect example of all three), it will not work efficiently. Ideally, a chiller should be placed in a separate room adjacent to the one housing the display aquarium. A cool place, such as a basement or garage, is best.

Right: Aquarium chillers use similar machinery to refrigerators and air conditioning units. They can be very effective, but are expensive to buy and run, and must be situated away from the tank.

Below: Most corals are adversely affected by very high water temperatures, so it is a good idea to protect a reef aquarium from overheating. A fan can cool the lights.

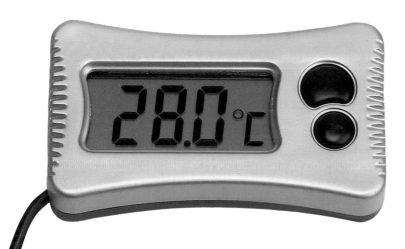

Left: Electronic thermometers, although more expensive than other types, are ideal for the reef aquarium. The probe is the only part of the thermometer that needs to be in the tank, and it can be positioned discreetly.

Lighting the aquarium

A key element of a reef aquarium is lighting of both the right intensity and the right type, because many corals and other invertebrates need light to live and grow. Often, the lights represent the single biggest investment, in both purchase and running costs, of the whole aquarium. To understand how to light the tank, we must consider the behavior of light underwater.

Light, water, and color temperature

Sunlight looks white to us, but consists of a wide range of light wavelengths, which are revealed as a spectrum of colors when it is passed through a prism, or in a rainbow. When light passes through water, different wavelengths (and hence different colors) are filtered out at different depths. Blue and violet light penetrate to a greater depth than red and orange light. This means that not only does the light become dimmer with increasing depth, but the color changes, too. Even in clear water, red and orange light are gone only 16 feet down, yellow light by 33 feet, and green light by 52 feet, leaving only blue and violet (so at this depth everything looks blue). In summary, we can say that the deeper we go underwater, the bluer the light.

This bluish light is what the zooxanthellae in corals and other invertebrates are adapted to use for photosynthesis, and so this is what lamps for reef aquariums need to produce. An indication of the range of wavelengths in light that looks white is given by the color temperature of a lamp. Color temperature, measured in degrees Kelvin (K), refers to the color of light emitted by an object when heated

WHAT IS COLOR TEMPERATURE?

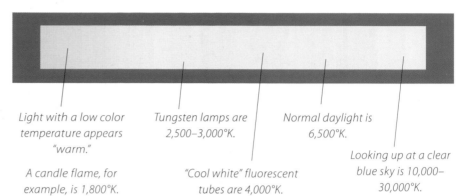

Light with a low color temperature appears "warm."

A candle flame, for example, is 1,800°K.

Tungsten lamps are 2,500–3,000°K.

"Cool white" fluorescent tubes are 4,000°K.

Normal daylight is 6,500°K.

Looking up at a clear blue sky is 10,000–30,000°K.

Right: *The shorter the wavelength, the more energy light has and the deeper it reaches. Corals are adapted to use those wavelengths that reach them at the depths at which they live. Reef tank lighting replicates this.*

to a certain temperature. The hotter the object, the bluer the light emitted (i.e., the higher the color temperature).

Sunlight has a color temperature of about 5,500K. The lamps usually used for reef aquariums have color temperatures of between 6,500K and 14,000K so, as required, their light is bluer than sunlight (but still looks white to us). 20,000K halide lamps are also sometimes used and these produce distinctly bluish light. In addition to these essentially "white" lamps, reef aquarium keepers often use "actinic" fluorescent tubes. These emit a narrow spectrum of blue-purple light (and some UV light). They are not essential if high color temperature "white" lamps are used, but they do make many

CHOOSING COLOR TEMPERATURE

The range of color temperatures available for both metal-halide and fluorescent lamps is wide, and which is best is a constant topic of discussion among reefkeepers. In practice, however, there seems to be little practical difference in how well corals grow under lamps ranging from 6,500K right up to 14,000K. You might find that you like the look of one type of lamp more than others, but this is the only real difference between them.

corals fluoresce, which can look very spectacular, and can be used for "dawn and dusk" transitions at the beginning and end of the day.

To mimic day length in the tropics, the aquarium should be illuminated for 12 to 14 hours a day. Light intensity can be varied during this period; six hours of lighting at full intensity is usually adequate, but longer is fine if this does not lead to overheating.

Types of lamp

Two types of lamps are currently used to provide lighting for reef aquariums. These are metal-halide lamps (a type of high-intensity discharge lamp) and fluorescent tubes of various types. Both have advantages and disadvantages.

Metal-halide lamps

Metal-halide lamps are the most popular lights for reef aquariums. They produce intense light with a visual effect much like sunlight. Metal-halide lamps run very hot, so be sure

Above: *Many corals have pigments that fluoresce when exposed to the blue-violet light produced by actinic lamps, making the tank look spectacular when these lamps are used alone, for example at "dawn" and "dusk."*

to position them about 10–12 inches above the water surface. They are usually used in pendant fittings with built-in reflectors, housing one, two, or three bulbs, often with built-in timers and sometimes additional fluorescent tubes. Wall-mounted metal-halide

lamps are also available. The bulbs usually used for reef aquariums have color temperatures of 6,500K, 10,000K, 14,000K, or 20,000K, rated at 70, 150, 250, and 400 watts. Some metal-halide bulbs emit significant amounts of UV light; always use the UV shields on lamp fixtures.

Most metal-halide fixtures will light a tank width of about 24–29 inches per bulb (although you can light a 35-inch wide tank reasonably well with a single lamp). Thus a 47-inch tank needs a

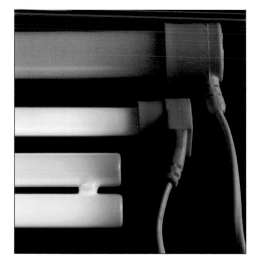

Above: *The standard fluorescent T8 tube (top) is available in various colors. T5 tubes (center) provide higher output for their size; also available in compact format (bottom).*

HOW MANY FLUORESCENT TUBES?

The rule of thumb is at least 40 watts of T8 tubes per 140 square inches surface area.

- For a 48 × 18 × 20 inch tank, the surface area is 48 × 18 inches = 864 inches^2.

- 864 divided by 140 = 6.17

- 6 x 40 watts = 240 watts required in total.

two-lamp unit, and a 71-inch tank a triple lamp unit. For tanks up to 18 inches deep, 150-watt lamps are adequate, although some corals show better color under 250-watt bulbs. Tanks 24 inches or more deep need at least 250-watt bulbs if invertebrates with high light requirements are to be kept close to the bottom, and 400-watt bulbs may be even better in such tanks.

Fluorescent tubes

Three types of fluorescent lights are commonly used to light reef aquariums: T8 and T5 tubes, and Power Compacts. T8s are the "traditional" 1-inch-diameter, double-ended fluorescent tubes. They are cheap to buy, but their light output is relatively limited.

T5 tubes are also double-ended, but .63 inch in diameter and more powerful than T8s (1.8 x higher light output is claimed), but with a higher power consumption. They are more expensive than T8 units, but you need fewer to achieve the same light intensity.

Power compacts are single-ended lamps that are effectively a pair of slim fluorescent tubes side by side. They produce about double the light intensity of T5s for the same length of tube, but power consumption and heat output are higher.

With all fluorescent lamps, multiple tubes are needed to provide enough intensity for a reef aquarium. For every 140 square inches of tank surface area you need at least 40 watts-worth of T8 tubes, or 30–35 watts of T5 tubes, and

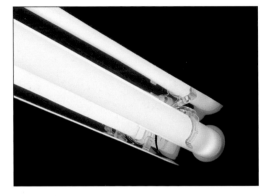

WHITE TRIPHOSPHOR FLUORESCENT TUBE

These curves show that this type of lamp produces light at a wide range of wavelengths, providing bright illumination for all creatures in the aquarium. (The vertical scale reflects comparative output. Wavelengths are in nanometers (nm)—billionths of a meter.)

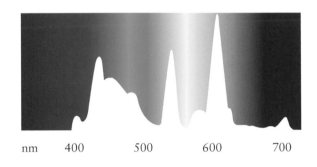

METAL-HALIDE LAMP

In common with the lamp above, this has a wide output with high levels at 400–480nm (good for zooxanthellae) and 550nm (to simulate sunlight). The volume of the curves here indicates a bright lamp.

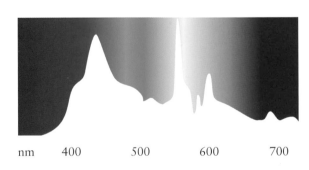

BLUE ACTINIC 03 FLUORESCENT TUBE

The output of this tube is concentrated in the blue area of the spectrum, especially in the so-called "actinic" range peaking at 420nm, which is vital for zooxanthellae to thrive.

It also supplies some UV for a fluorescent effect.

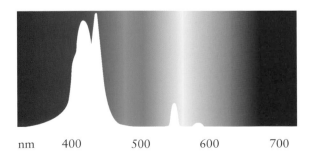

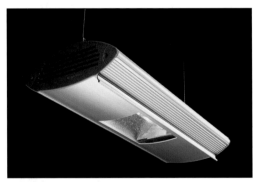

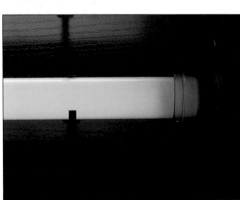

METAL-HALIDE LAMPS

Advantages

- Penetrate better into deeper water: in tanks 24 inches or more deep, even the more powerful fluorescents struggle to get enough light to the bottom.

- Cheaper to run than fluorescents for the same intensity of light.

- Simple wiring (one cable per unit).

Disadvantages

- Must be suspended above the tank.

- More expensive to buy.

- Heavy and bulky.

- Run very hot—heat output can be a major issue.

FLUORESCENT LAMPS

Advantages

- Run relatively cool.

- Cheaper to buy.

- Low profile—they can be fitted into aquarium hoods or slim luminaires.

Disadvantages

- Need multiple tubes, so the wiring required can end up as spaghetti-like tangles (unless luminaires are used).

- May need a custom-built hood to accommodate them.

- Starter units must be housed somewhere.

about 35 watts with Power Compacts. Things are slightly more complicated with Power Compacts, as the range of tube lengths is more limited; on longer tanks, pairs of tubes mounted end-to-end will span the whole length of the tank.

All three types of fluorescent lamps can be used in lighting hoods or in luminaires suspended above the tank, with the tubes connected to the power supply using water-resistant end caps. Fluorescent tubes need to be used with good reflectors to maximize the light reaching the aquarium. A mix of white tubes (6,500 to 14,000K) and blue "actinic" tubes is usually used; the ratio of white to blue tubes is not critical.

Choosing between fluorescents and metal-halides

Metal-halide and fluorescent lamps each have their advantages and disadvantages and these are summarized in the panels shown on this page. You should also bear in mind that the visual effects of fluorescents and metal-halides are very different; fluorescent tubes produce a soft, even light, whereas metal-halides provide a sharp, harsh light.

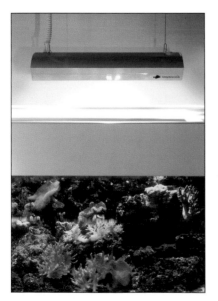

Left: Suspend metal-halide lamps 8–12 inches above the water to allow air circulation for cooling, to avoid splashes and to allow the light to cover a wide area of the tank.

Right: This fitting has metal-halide lamps, plus white and actinic tubes. The latter can light the tank when the full intensity from the metal-halides is not needed.

Biological filtration

Place a fish or other aquatic creature into an aquarium and almost immediately a range of waste products will begin to accumulate in the water. These nitrogen- and phosphorus-containing compounds are the by-products of metabolism. Solid wastes (which in the reef aquarium include not only fish feces, but also such things as uneaten food and the occasional dead animal) are eventually broken down by bacteria into similar chemical compounds.

These compounds are deleterious to the reef aquarium for two reasons. Firstly, they can be directly toxic to the aquarium inhabitants and secondly, they act as nutrients that stimulate the growth of algae, which is often undesirable in the reef aquarium. Such compounds are present in only minute

Below: When fishes and other marine animals are kept in the aquarium, their waste products accumulate in the water, and must be removed or processed before they reach toxic levels.

concentrations in the water around coral reefs and many reef animals are unable to tolerate high levels of them. Fortunately, methods of removing these compounds have been developed and these make it possible to keep aquariums without continuously changing the water. These methods are known as biological filtration.

Biological filtration— the traditional way

There are two main biological routes for dealing with waste products in the reef aquarium. In most tanks both operate in parallel, although their relative contributions vary. These are direct uptake of waste products by algae and biochemical processes mediated by bacteria. Traditionally, only the latter have been referred to as biological filtration, but in practice both processes are important. How algae use chemical waste products and how this process can be exploited to keep the aquarium water in good condition is discussed further on pages 50–52. Here, we look

more closely at bacterially mediated biological filtration.

Converting ammonia to nitrite

The primary substance dealt with by bacterial biological filtration is ammonia. This is excreted directly by fish and other aquatic creatures and is the end product of the breakdown of more complex waste products. Ammonia is highly toxic to marine animals.

In the classic biological filtration process, ammonia is taken up by certain types of bacteria and converted into a compound called nitrite. It was previously believed that only one type of bacteria *(Nitrosomonas)* could do this, but this proved not to be the case; many other bacteria can convert ammonia to nitrite, among them species of *Nitrosospira* and *Nitrosococcus*. Nitrite is less toxic to aquatic animals than ammonia, but is still far from harmless.

Converting nitrite to nitrate

Another group of bacteria can convert nitrite into nitrate, which is still less toxic. Some marine animals can tolerate quite high nitrate levels, although others are less hardy in this respect. These bacteria include species of *Nitrobacter, Nitrococcus, Nitrospina,* and *Nitrospira. Nitrobacter* species, which are usually quoted as performing this reaction, are actually not important in marine environments.

The whole process of ammonia being converted to nitrate (via nitrite) is known as nitrification. Nitrifying bacteria are ubiquitous in natural environments. These bacteria have certain requirements, most importantly

THE BIOLOGICAL FILTRATION PROCESSES IN A TRADITIONAL MARINE AQUARIUM

A piece of shrimp meat is thrown into a marine aquarium fitted with a traditional biological filter. In this system, whatever nutrients are not removed from a piece of food by a fish are either turned into nitrate, used for algal growth, or remain dissolved in the tank water.

The fish also excretes the waste products of its metabolism of the nutrients obtained from the shrimp. These are ammonia, which is dealt with by the biological filter as shown here, and various phosphorus-containing organic compounds. The latter cannot be handled by the traditional biological filter (which only really deals with nitrogenous wastes), but algae use them as nutrients for growth.

It is digested by the fish, which extracts the nutrition it can from the shrimp, and uses these nutrients for its growth and metabolism.

The shrimp is eaten by one of the fish in the aquarium.

Bacteria in the filter convert ammonia to the less harmful (though still toxic) nitrite.

AMMONIA

The undigested remains of the shrimp are excreted as feces, and drop to the bottom of the tank. Bacteria rot these down, using what they can for their own growth, and producing ammonia.

BIOLOGICAL FILTER

Nitrate accumulates in the aquarium water.

Algae will also pick up some of the ammonia, but if a very efficient biological filter is present, they will not be able to compete very well with the bacteria and will get only a small share, although they may also use the nitrate produced by the biological filter

NITRITE

NITRATE

A different population of bacteria convert the nitrite to nitrate, which is even less harmful to fishes.

35

a solid surface to colonize and a supply of oxygen. The chemical reactions of nitrification involve sequential addition of oxygen to the ammonia molecule to produce nitrate.

Removing nitrates

Nitrate is much less toxic than ammonia, but high nitrate levels in reef aquariums have been associated with a variety of adverse effects. These range from the failure to thrive of a variety of marine fish, through excessive growth of algae, to poor growth and coloration in corals. In a reef aquarium, it is therefore necessary to remove as much of the nitrate as possible.

There are a number of ways to do this. Algae can be deliberately grown to absorb the nitrate and then harvested (see page 50), special media can be used in filters, or you can perform very frequent partial water changes. One very convenient way to remove nitrate is provided by yet more bacteria, which acting in concert can convert nitrate to nitrogen gas, which is then lost to the atmosphere. This reaction is known as denitrification. In chemical terms, it involves removing the oxygen from nitrate, the opposite (as its name suggests) of nitrification.

The bacteria that perform these reactions include a wide range of species. Important ones in the marine environment include members of the genera *Flavobacterium*, *Pseudomonas*, *Vibrio*, and *Aeromonas*. What these bacteria have in common is that they are all members of a group known as facultative anaerobes, which means that they can grow well with both high oxygen levels and where there is no oxygen available. Denitrification, however, happens only at low oxygen

levels. If they cannot get oxygen from more straightforward sources, the bacteria take up nitrate (or indeed nitrite or nitrous oxide), and take the oxygen that is chemically bound in these substances.

Combining nitrification and denitrification

Nitrification and denitrification require very different conditions, chiefly with respect to oxygen levels. Providing the ideal conditions for nitrification to occur has been the driving force in the development of biological filtration systems for both freshwater and marine aquariums. Many such systems have been developed and they are generally very effective in converting ammonia to nitrate. However, in reef aquariums it is important to keep nitrate levels low, and this makes it difficult to rely on traditional biological filtration systems.

The ingenuity of reef aquarium enthusiasts led to the development of a variety of technological systems to provide the right low-oxygen conditions

Above: A thriving reef aquarium shows the beauty of live rock as well as its efficiency as a biological filter. This mature aquarium plays host to a range of corals and fish.

for denitrification, while running on aquariums with the very high oxygen levels needed for their inhabitants and for the nitrification side of the biological filtration process. Such systems have often been complex and difficult to set up and run. However, as reef aquarium techniques have evolved, very simple systems have been developed that do exactly what complex technological solutions often failed to do in terms of creating conditions that favor both nitrification and denitrification, as well as providing a host of other benefits. The key ingredient in these systems is live rock.

Live rock—coupling nitrification and denitrification

Live rock, also called living rock, is at the heart of most modern reef aquariums. Of course, it is not literally alive. Instead,

BIOLOGICAL FILTRATION PROCESSES IN A REEF AQUARIUM

The end result of the complex ecosystem in a tank with live rock is that by the time any nutrient input has passed through this whole web of animals and algae, the nitrifying and denitrifying bacteria tend to have less of a biological load to handle. Together with the almost unique coupling of bacterial nitrification and denitrification, this makes live rock-based systems tremendously efficient biological filters.

Nitrogen gas is lost to the atmosphere, completely removing nitrogenous wastes from the aquarium.

As the algae grow, they are grazed by a large number of small creatures living on and in the rock.

The fish eats the shrimp, and extracts what nutrients it can from it.

The ammonia and phosphorus-containing compounds excreted by the whole chain of animals that participated in the consumption of the piece of shrimp are taken up by algae growing on the live rock.

Nutrient input as fish food—frozen shrimp.

The feces of the fish fall to the bottom of the tank, but instead of remaining there to rot, they are quickly eaten by a brittle star living among the rocks.

Bacteria on and within live rock convert ammonia to nitrate, and nitrate to nitrogen gas.

Next, tiny crustaceans consume the feces of the brittle star, and extract what nutrients they can, and so on.

The feces of the brittle star, however, contain fewer nutrients than those of the fish. Passage through another animal has led to more of the nutrients being removed from the original piece of shrimp.

This brittle star extracts what nutrients it can from the feces, before excreting what it cannot digest.

Above: When broken up by waves and storms, rock like this on a coral reef can be harvested as live rock for the aquarium. Note the variety of shapes and sizes.

Below: Collecting live rock in the wild. This large, attractive piece will command a high price. Air transportation makes up a high percentage of the final retail value.

it is material with many living organisms on and in it. It can be colonized by a vast range of animal and plant life, but most importantly from the perspective of biological filtration, it harbors large populations of nitrifying and denitrifying bacteria.

Live rock has a very porous structure that allows water to percolate slowly through it. This structure not only provides a huge internal and external surface area for bacterial growth, but also allows zones with high and low oxygen concentrations to exist in close proximity. This is what permits both nitrification and denitrification to occur within the same piece of substrate.

Oxygen levels in the water are high at the surface of the rock, and this area holds a large population of nitrifying bacteria. As part of the nitrification process, these bacteria deplete much of the oxygen from the water as it flows over them, but deeper within the rock, denitrifying bacteria thrive in this deoxygenated water and reduce the nitrate to nitrogen gas. Live rock can thus act as a complete biological filter, converting ammonia to nitrate, and nitrate to nitrogen gas.

Live rock—harnessing biodiversity

This in itself would be an impressive performance. But live rock can do much more. As we have seen, live rock is colonized by a vast range of algae and animals. Some of these are obvious and attractive and they make a great contribution to the aesthetics of a reef aquarium. They include the pink and purple coralline algae that encrust the surface of good-quality rock, the tiny fanworms that spread their feathery tentacles from the mouths of their tubes, the interesting sponges, polyps, and juvenile corals that may appear.

Less obvious, but in some ways more important, are a myriad of tiny worms, crustaceans, brittle stars, and so on. What the presence of these creatures does is to mimic the situation on a wild reef. As discussed on pages 10–11, one of the key features of coral reefs is that they are very nutrient-efficient. If any type of nutrient becomes available, it is quickly taken up by some form of

Below: Boreholes like these are made by a variety of marine animals. They weaken coral rock and increase the chances of pieces breaking away during storms.

Above: Unpacking cured live rock as it arrives from the importer at an aquarium dealer's shop. The dealer will transfer it to a holding tank until it is sold.

life and the huge diversity of animals and plants on a reef means that every possible source of nutrients is exploited.

If we transplant some of that biodiversity into the aquarium, which is effectively what happens when we add live rock, we can use the biodiversity to help us deal with the waste products. The more complex the aquarium ecosystem, the more effectively nutrients (in the form of food added to the tank or the waste products of animals) are handled.

Live rock, types and origins

Any rock that is left in the sea in the vicinity of a coral reef will eventually (over a period of months and years) become "live" as it is colonized by a range of algae and animals. In fact, "farmed" live rock, made by leaving either quarried rock or specially made concrete in the oceans for some time, is sometimes available. However, for the most part, live rock is generated naturally and consists of dead coral skeletons or chunks of dead coralline algae, sometimes partly fossilized. Having been weakened by the activities of worms, mollusks, sponges, and other animals that bore into the reef structure, pieces are broken off the reef by a combination of wave action or storms. This material (essentially reef rubble) is typically deposited in lagoons, from where it is collected and shipped to the aquarium industry.

Although live rock collection has been the subject of restrictions in some areas, because of concerns about its environmental impact, it appears to be a sustainable practice. Properly collected live rock is not formed over a long geological time period, nor is it an integral part of the living reef. The stony corals on the reef are subject to a continuous process of death and

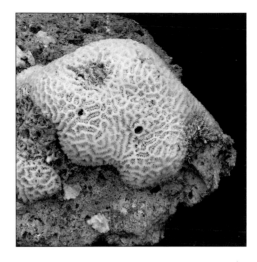

Above: In the reef aquarium, inert material, such as this dried coral rock (sold as "coral bones"), is rapidly colonized by bacteria and a range of other marine life.

recolonization (which is how the reef is built in the first place) and most living rock consists of the dead skeletons of fast-growing corals such as *Porites, Montipora,* and *Acropora* species (see pages 71–73) or of coralline algae. These species can grow very quickly and substantial pieces of living rock may represent only a few years of growth. Live rock left in the lagoon would probably be turned quite quickly into sand and gravel by the same types of tunnelling creatures that helped to cause it to be broken off the reef.

Live rock differs in character depending on its place of origin. The types of corals or algae that constructed the rock vary from place to place, as do the types of animals and plants that subsequently colonize it. The most commonly used types of live rock are sourced from Fiji and Indonesia (the former also providing a great deal of artificial living rock, made from a special concrete mix). These are typically very porous and quite light in weight for

Right: Most live rock is primarily made up of the skeletons of fast-growing stony corals and other reef organisms, as can be seen in this piece.

their volume. Rock (whether natural or artificial) from Fiji is often covered with extensive growths of attractive plum-purple coralline algae, as well as algae of other colors and types.

Live rock is also available from a range of other areas (although in some cases only sporadically); rock from each location has its own characteristics. Rock is usually sold according to either its geographical origins (Fiji or Indonesian, for example) or the characteristic shapes of the rock (examples are plating rock or branch rock), or sometimes both (Tongan branch rock is one example).

Not all live rock has the same amount of desirable or attractive life on it, and relatively unattractive rock, with little or no coralline algae, is frequently sold at lower prices than more colorful

material, and is often called "base rock." This type of rock remains very useful in filtration, with a similar bacterial population to more attractive material and just as many small crustaceans, worms, and so on. In time, in the right tank conditions, this rock will become covered in coralline algae and be colonized by more visible invertebrates, such as sponges and tubeworms. Base rock provides a useful, inexpensive form of live rock for use in the less conspicuous parts of the aquarium.

Transporting live rock

Shipping live rock can have drastic effects on the life growing on it, as can the way it is handled on receipt. Because the price of transporting the rock depends on its weight, it is usually

Below: *This particularly attractive piece of live rock is covered with a range of coralline and other algae, as well as some mushroom anemones. It will make an instant impact.*

shipped wrapped in wet newspaper rather than in water. Not all the life on live rock can survive transportation like this and the longer the journey, the more that is likely to be lost. Rock that has been subjected to a relatively short flight (for example from the Caribbean to the United States, from East Africa or the Red Sea to Europe) is likely to be in better condition, in the sense of having more of its original fauna and flora surviving, than rock shipped from, say, Indonesia to either the United States or Europe. Reef aquarium keepers living in regions that are close to reef areas may be able to obtain raw rock in excellent condition because of the short transport time.

Curing live rock

Live rock also needs careful handling when it arrives at its destination. Anything that dies in transit will immediately begin to rot. The rock needs to be carefully cleaned of any large deposits of decaying material

WHAT YOU MIGHT FIND ON A PIECE OF LIVE ROCK

Live rock in detail. This piece of Fiji rock shows a number of typical features of this material. When you are buying live rock, look carefully at each piece to see what it has to offer.

Sponges often grow on live rock; they tend to die off during shipping and curing, but grow back afterwards.

and then housed in a sufficient volume of water with adequate filtration capacity to handle the heavy load of decaying organic matter that cannot easily be cleaned off. The idea is to control ammonia levels (which can be extremely high when a great deal of decaying organic matter is present) such that the remaining life on the rock is not poisoned.

It is possible to buy rock in this raw, freshly imported condition, but it requires very careful handling when used to set up a reef aquarium. As well as producing very high ammonia levels when added to the aquarium, raw rock also produces considerable quantities of detritus. If this is not removed, it can lead to algae problems later in the tank's life.

When raw live rock is used to set up a reef aquarium, it must be placed into a tank with strong water movement and, preferably, with a powerful skimmer. The ammonia and nitrite levels of the water must be monitored carefully (as when maturing a traditional biological filter) until both have fallen to zero. In addition, detritus must be siphoned away as it forms, and it is wise to perform extensive partial water changes in order to prevent an accumulation of nitrate, as the quantities produced during this process can easily overwhelm the ability of the rock to remove it. The tank is only ready to stock once ammonia and nitrite

Grey-black areas may indicate patches of anaerobic decay, suggesting the rock may not be fully cured.

Worms are very common inhabitants of live rock; many are resilient and survive shipping and curing, and are very useful aquarium inhabitants.

Coralline algae of various types are the most obvious and colorful inhabitants of good-quality live rock.

levels have fallen to zero and detritus production has ceased.

This process is variously referred to as "curing" or "seeding" the live rock. Once it has been completed the live rock is a fully matured biological filter. However,

the process itself is an unpleasant one to carry out in the home; the tank water may become very turbid, due to the high levels of bacteria, and foul-smelling. This passes eventually, but the process may take several weeks.

MATURING AQUARIUMS

Traditional biological filter

When using a traditional biological filter, a very important step in setting up the aquarium is maturing the filter. This is essentially the process of ensuring that the filter media are colonized by sufficient numbers of nitrifying bacteria to handle the quantities of waste that will be produced by the aquarium inhabitants. The usual method is to provide a source of ammonia (either by adding ammonium chloride or a similar compound to the water, or by allowing a small quantity of fish food to decay in the tank) and a source of nitrifying bacteria. The latter could be some sand or gravel from an established aquarium, a commercially available bacterial starter culture—or nothing at all. Nitrifying bacteria are very common in the environment and will find their way into the aquarium without any intervention from the fishkeeper, although this approach makes the process much slower. Having added both ammonia and bacteria, the next step is to wait and to keep testing the water for both ammonia and nitrite. The ammonia level will be high at first, but will decline as the bacterial population increases; as this happens the nitrite level will rise, often peaking as the ammonia level hits zero. Eventually, the nitrite level will fall to zero, and the filter is mature and the tank is ready to stock.

Using live rock

When using live rock as a biological filter, this traditional maturation process is bypassed. When cured rock is used to set up an aquarium, this is like transplanting a fully mature biological filter into the tank. All that is necessary after adding the rock is to check ammonia and nitrite levels for a few days to check that the rock is properly cured. If so, ammonia and nitrite will be undetectable, and the tank is ready to be stocked. If traces of either ammonia or nitrite are present, either the rock was not quite completely cured, or something died in transit between the dealer and your aquarium. Waiting a few days and rechecking ammonia and nitrite to make sure that they are zero is usually all that is required.

Above: *Fully cured live rock is not only attractive, but also forms a ready-matured biological filter for the reef aquarium.*

Fortunately, it is not necessary to cure live rock at home. Many importers and larger marine dealers sell cured live rock, having done all the hard work. Large quantities of freshly imported rock can be processed in large vats of saltwater, often equipped with very large skimmers and powerful pumps to circulate the water, before being offered for sale. Cured live rock is more expensive than raw rock, but has the great advantage that it is ready to use as a biological filter almost immediately. It requires little special handling, so the start-up time for a new aquarium is greatly reduced. Cured rock is less smelly, too.

Another advantage of cured rock is that more of the original life on the rock is likely to have survived after processing by a good dealer or importer. Very large curing systems, often containing several thousands of gallons of water, help to keep the ammonia levels down, and keep casualties to a minimum.

In a home aquarium, it is quite likely that the ammonia levels will be very high for a few days as a result of the relatively small volume of water in the system. This ammonia peak will be followed by a high nitrite level before the rock eventually settles down and water conditions stabilize. During this period, there is a high probability that more of the life on the rock will succumb. This is not to say that everything will be lost if you set up your aquarium with raw rock. Many of the plants and animals that grow on live rock are very resilient and will survive even such exacting conditions. Similarly, there will be no adverse effects on the nitrifying and denitrifying bacteria. It is only the more delicate animals, such as juvenile soft and stony corals, that are at risk, but preserving them may be well worth the extra expense of buying cured rock, especially when combined with the convenience of using it to set up a tank.

Buying live rock

When looking for live rock, it pays to shop around. Different retailers stock rock of different types from a variety of locations, and it is best to visit several aquarium dealers to take a look at the available selection and see how it fits with your idea of the way you want your tank to look. Another thing to consider is the size of pieces of rock that are available. Large pieces (although often harder to find) can make constructing attractive tank layouts easier, although they can be more difficult to handle and are not ideal for small tanks. Prices may also vary considerably from one dealer to another, even for rock of similar quality from the same location, and as live rock is one of the most expensive items in the whole reef aquarium system, significant savings can be made if you buy from the right source.

DON'T MIX FILTERS

When using live rock, it is important not to use any other kind of biological filtration, particularly not highly efficient nitrifying filters such as trickle filters. Using these in addition to live rock often results in high nitrate levels in the aquarium water, and there is no need for supplementary biological filtration if an adequate quantity of live rock is present.

Having chosen a type of rock that you like, the next thing to do is to check that it is properly cured. There is a very simple way to do this—the smell test. Properly cured live rock is relatively odorless, at worst, there is a slight scent of seaweed or fresh seafood. If the rock is not quite cured it smells terrible. In most cases, all that is required is longer in the curing tank, so if you find some rock that you particularly like but it is still smelly, the best idea is to ask the dealer if they will keep the rock a little longer if you pay a deposit on it.

Sand beds

A key addition to the tank's biological filtration system is the bottom substrate: sand and/or gravel. Sand has a huge surface area for bacteria to colonize, and a sand bed with well-oxygenated

Above: Sniffing live rock is a good way to detect any decaying material. Make sure you smell each piece before buying and avoid any chunks that need further curing.

water above it creates a gradient of oxygen concentration that allows both nitrifying and denitrifying bacteria to coexist within the bed. This can increase the efficiency of biological filtration, providing a great adjunct to live rock.

Choosing sand

A wide range of sands and gravels are suitable for reef aquarium use. The most commonly used are calcareous sands, i.e., those composed of calcium carbonate (usually in the form of aragonite). Most of them are ultimately derived from coral (although other suitable calcareous sands are available).

Right: Most sands and gravels used in marine aquariums are ultimately derived from stony coral skeletons, and are white or ivory-colored.

Non-calcareous sands can be used in reef aquariums, provided they have no metal content, although they do not make the contribution to the maintenance of calcium and alkalinity levels that aragonite-based sands do.

Suitable sands are available in a range of different grades, from coarse gravel to very fine sand, with particle sizes ranging from under 0.1mm to about 10mm. Material in the middle of this range (0.5–5mm) is most often used in the reef aquarium, but finer and coarser substrates have their roles. The particle size is important in biological filtration because it determines how

Below: A range of different marine sands are available pre-packed in "live" form. These are very convenient to use and help with biological filtration.

deep a bed is required to allow both nitrifying and denitrifying bacteria to grow. You can create a shallow bed using fine sands. A one-inch bed made up of sugar-fine sands (with a grain size of about 0.5mm) or very fine sands (with 0.1mm grains) can be effective. This allows the sand bed to be very

unobtrusive and take up little tank space. If you use coarser sand, you will need a deeper bed for denitrification to take place.

The vigorous water movement employed in reef aquariums may lead to fine sand being blown away from spots where the current is particularly strong. Patches of coarser sand or gravel can be used in these areas.

If the bed is mainly made up of fine sand, you can add a proportion of coarser material of different grades, right up to 10mm pieces of coral gravel. This has no adverse impact on filtration and can help the sand bed fulfill many different needs for the aquarium inhabitants. The fine sand is good for

burrowing invertebrates, such as worms and sea cucumbers, for fish that feed by sifting sand, and for those that bury themselves to sleep. The presence of coarser material is useful for creatures such as pistol shrimp which construct tunnels within the substrate.

Live versus inert sand

Sand is available in both inert (dry) and live forms. The most widely available live sands are sold ready-colonized by a variety of nitrifying and denitrifying bacteria, packed wet in sea water with additives that preserve the bacteria. A range of different types of sand are available in this form. In addition, sand collected from the ocean in the tropics and shipped wet to keep alive the resident worms, crustaceans, and other invertebrates is occasionally available. Live sands of both types can be added straight to the aquarium after draining off the water they are packed in—no washing is needed.

Inert sands are available in a wider range of types than live sand, and are considerably less expensive. Wash them before use to remove fine dust and any contaminating material, such as bits of driftwood, shreds of plastic, and seaweed stalks. In an aquarium with live rock, inert sand will rapidly become "live," and live and inert sands can be mixed.

Above: *The yellow-headed jawfish (*Opistognathus aurifrons) *needs a deep gravel bed to build a burrow. Many fishes and other creatures need a mix of fine and coarse material to build burrows.*

Left: *A sand bed completes the appearance of a reef aquarium, as well as contributing to the biological filtration and providing a habitat for a range of creatures.*

Protein skimming

Protein skimmers, or foam fractionators, as they are more accurately but less frequently known, provide a valuable component of aquarium filtration. An efficient skimmer can remove 80% or more of the organic waste produced in the tank, reducing the load on other filtration methods.

Skimmers work because many organic molecules, including many waste products of marine life, are attracted to air-water interfaces—as can be seen in the films that build up on the surface of tanks with little water movement. Skimmers exploit this effect by injecting fine air bubbles into a column through which tank water is flowing. The bubbles provide a large surface area to bind organic molecules, and when this happens, a thick foam is formed. This rises up through the skimmer to the top of the column, where it collapses into concentrated liquid waste that is collected in a cup, while the water is returned to the aquarium.

Skimmers all operate according to this basic principle, but different designs use a variety of ways to generate fine bubbles, and to keep the bubbles in contact with the water for as long as possible (which increases the efficiency of skimming).

Types of skimmer

There are many different skimmers on the market, designed for tanks of all sizes from very small to very large. There are skimmers designed to be fitted inside the aquarium, to hang on the outside of the tank, to fit in a sump, or to be completely free-standing. Although there are other designs, reef aquariums usually use either venturi skimmers or needle-wheel (or aspirator)

ANATOMY OF A PROTEIN SKIMMER

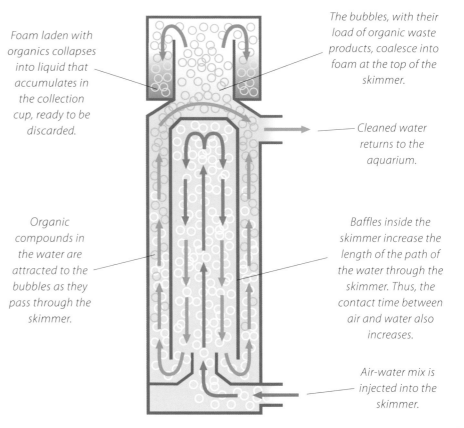

Foam laden with organics collapses into liquid that accumulates in the collection cup, ready to be discarded.

The bubbles, with their load of organic waste products, coalesce into foam at the top of the skimmer.

Cleaned water returns to the aquarium.

Organic compounds in the water are attracted to the bubbles as they pass through the skimmer.

Baffles inside the skimmer increase the length of the path of the water through the skimmer. Thus, the contact time between air and water also increases.

Air-water mix is injected into the skimmer.

skimmers. Both types are easy to install, require little adjustment or maintenance and are highly efficient. Simple skimmers that use wooden air diffusers are useful in some circumstances, such as in small quarantine or hospital tanks, but are seldom used in display tanks.

Venturi skimmers create very fine bubbles using a venturi valve, which sucks air into a high-pressure stream of water, breaking it up into fine bubbles in the process. This is a very good method of generating foam, and venturi skimmers are very effective. These skimmers require powerful pumps to drive them, as venturi valves require water to be passing through them at high pressure in order to

work efficiently. Such pumps can be expensive to buy and to run (in terms of electricity consumption) and they can produce considerable quantities of heat, which in some systems can

Above: The surfaces of the bubbles in a skimmer attract organic molecules, allowing them to be removed from the water.

Left: *Foam accumulates at the top of a protein skimmer, then collapses into a thick, dark (and often smelly) brown liquid, rich in organic waste products, algae, and bacteria.*

contribute to overheating of the aquarium. Also, venturi skimmers seem to be particularly sensitive to very minor fluctuations in water chemistry; adding some trace element supplements can cause venturi skimmers suddenly to start producing large quantities of wet foam, potentially leading to wet carpets. On the other hand, these skimmers will often foam less effectively shortly after you have been working in the tank, a phenomenon that is believed to be related to tiny quantities of natural oils being shed from the skin and affecting the surface tension of the aquarium water. These things happen with all skimmers, but they seem to be more pronounced with venturi models.

Needle-wheel skimmers work rather like an egg whisk, beating air and water together to produce foam. The heart of the skimmer is a rapidly spinning rotor, a modified pump impeller covered in pins or blades. Water and air are passed over the rotor, which mixes them rapidly, "chopping" a lot of air into the water to produce very fine bubbles. Needle-wheel skimmers produce a very dry foam, which extracts the maximum amount of organic waste for a minimal water loss. They are extremely efficient, often very compact skimmers, outperforming most venturi models of comparable or even larger size. Needle-wheel skimmers have an additional advantage in that they do not require such powerful pumps to drive them. A small, relatively low-wattage motor powers the rotor, in some cases also pushing water through the skimmer. In other designs (usually larger skimmers) another pump (which again does not need to be as powerful as one driving a venturi skimmer) provides the water flow. This makes needle-wheel skimmers more economical to run and less likely to contribute much excess heat to the tank.

To skim or not to skim?

Skimmers are not without their disadvantages. As well as removing dissolved organic waste products they can, in theory, remove trace elements—particularly iodine—from the aquarium water. However, this does not seem to be a real problem in practice and if it does happen, it is easy to replenish trace elements using supplements.

More of an issue is that it is not just organic molecules that adhere to air-water interfaces, such as the surfaces of the bubbles in skimmers. Small suspended particles also do so, and are thus removed from the water by skimming. While this is good in some ways (removing suspended detritus, for example), some of the fine particulate matter suspended in aquarium water is beneficial. Bacteria, microscopic algae,

Above: *Internal baffles in this protein skimmer extend the path the water-air mix takes, thus increasing the contact time and making it as efficient as a much taller model.*

Left: The water around coral reefs has a very low level of organic pollutants. Protein skimmers help to mimic such conditions in the reef aquarium.

and larger forms of plankton can all be present in quite large quantities in the water of a reef aquarium. A wide range of filter-feeding organisms (including corals) benefit from the smaller plankton, and planktivorous fish feed on the larger organisms.

Very powerful skimmers can strip much of this plankton from the tank water, although this process does not affect all types of plankton equally. Smaller plankton, including bacteria and microscopic algae, are more efficiently removed than larger plankton, such as small crustaceans and other zooplankton. In part, this is because larger reef plankton tend not to be evenly distributed in the tank water, waiting to be sucked into a skimmer; rather, they tend to be closely associated with solid substrates. However, filter-feeding organisms, such as sponges and fanworms, do grow more prolifically in tanks without skimmers and corals often show better polyp extension.

However, this must be balanced against the valuable role that skimmers can play in keeping down organic wastes in the aquarium, particularly if the tank is home to many fishes. The lack of plankton can be compensated for by feeding the tank well or by using a refugium (see page 53). Although the majority of reef aquarium keepers use skimmers, there are alternatives, typically based on using algae to absorb wastes, and these systems are claimed to maintain larger populations of plankton.

NEEDLE-WHEEL PROTEIN SKIMMER

This skimmer can be fitted in a sump or hang on a sump or display tank.

Collection cup in which the waste-laden foam collapses.

A pipe can be connected to this port to drain off the waste.

Cleaned water returns to the tank over this cascade.

Optional basket for chemical or biological filtration media.

Reaction chamber in which the water and air bubbles remain in close contact.

Optional surface skimmer attachment on the inlet pipe.

Pump with an 18-blade impeller that generates fine air bubbles in the incoming air/water flow.

Selecting a skimmer
Several factors will influence the choice of a skimmer. The first is tank size; manufacturers rate their skimmers

TRIPLE-PASS PROTEIN SKIMMER

This is the skimmer shown at the bottom of page 47, with the pump and pipes connected for use in hang-on mode.

The skimmer has a large collection cup, so should require infrequent emptying.

Water returns to the tank or sump through these twin outlet pipes.

The needle-wheel pump sits inside the tank or sump and injects the air-water mixture into the skimmer.

according to the size of tank they are intended to serve. It is worth noting that such ratings tend to be optimistic, and it is often a good idea to choose a skimmer rated for a slightly larger tank than you have. The second consideration is whether the aquarium has a sump; sumps will accommodate a wide range of skimmers. If it does not, you must use an in-tank or hang-on model. Bear in mind how much headroom there is in the cabinet (if you are using one) and under the tank, and think about price—many skimmers are quite expensive. Fortunately, there is a huge range of skimmers on the market and at least one to suit every reef aquarium.

HANG-ON PROTEIN SKIMMER

Above: *This protein skimmer is designed to hang on the display tank, with only the inlet and outlet pipes inside the aquarium.*

Right: *The rear of the skimmer, as seen on the outside of the aquarium, is sleek and unobtrusive.*

Algae filters and refugiums

Biological filtration in aquariums has long been based primarily on bacteria processing nitrogenous waste products. In marine aquariums this has been supplemented by skimmers, which reduce the burden on biological filters by removing a wide range of waste products. The use of living rock shifts the process closer to that in the ocean by introducing algae as part of the system.

The role of algae

Algae take up both nitrogenous waste products and phosphates from the water around them and use them as nutrients to fuel growth, in much the same way that terrestrial plants use similar materials as fertilizers. On the reef, algae play a tremendously important role in the cycling of nutrients and this is increasingly being reflected in methods for keeping marine aquariums. This is shifting the way that reef aquarium enthusiasts view algae; no longer are they seen just as a nuisance.

Algal turf scrubbers

Many public aquariums use filtration systems based on algal turf scrubbers. These use turfs of filamentous algae growing on screens as the primary method of removing organic waste products, nitrogenous and otherwise, from the system. The algal turfs are cropped regularly, which completely removes nutrients from the aquarium.

Macroalgae

Although effective, for various reasons algal turf scrubbers have never really caught on in home aquariums. In this setting, filters based on macroalgae

Above: *Seagrass beds act as natural refugiums in the wild, being home to a huge range of small invertebrates and juvenile reef fishes.*

(seaweeds) are more popular. These provide very effective biological filtration, can be compact and low maintenance and can allow the aquarium to be run without a skimmer. They have the additional advantage that they can produce considerable quantities of planktonic organisms for corals and fishes to feed on. They do this by acting as a refugium, a part of the system that predatory fishes cannot reach, and where small worms, crustaceans, and other organisms (collectively known as microfauna) can reproduce freely, grazing on the algae.

Using a sump

To use macroalgae for filtration, the tank water needs to be circulated through a sump that is equipped with lighting and has a substrate to which the algae

can attach. The lighting can be kept on 24 hours a day or can be set up so that it is on when the main tank lighting is off. The latter method is known as reverse daylight photosynthesis. Photosynthesizing algae produces oxygen and consumes carbon dioxide, and both methods keep oxygen and pH levels high 24 hours a day, avoiding the drop in both that typically occurs at night in the reef aquarium. The macroalgae used for filtration generally do not require very intense light and one or two standard fluorescent tubes over the sump will provide enough illumination.

The Ecosystem method

The use of macroalgae for filtration was refined by American enthusiast Leng Sy to create a technique known as the Ecosystem method. It uses no skimmer or carbon and apart from some living rock in the main aquarium, relies for filtration entirely on a refugium filled with *Caulerpa* or other fast-growing

Right: Algae are a hugely important part of the reef ecosystem but are not always obvious, because of the intense grazing pressure, from tangs, for example.

algae. This method defines the size of refugium required for a given size of tank, uses a special mud substrate for the algae to attach to (which is claimed to supply all the trace elements required by the aquarium) and requires continuous lighting and regular harvesting of macroalgae.

This methodology has proven to be very effective for biological filtration, although whether it is better than similar systems remains to be established. Many aquarists who have used this system also report particularly good expansion of coral polyps, indicating that a lot of plankton is produced.

Some enthusiasts for algae filters suggest that the quantity of living rock in the aquarium should be reduced, as the algae take up much of the biological filtration load. It is possible to do this

Above: Caulerpa species are ideal algae for a refugium, as they grow fast and take up a lot of nutrients from the water.

A MACROALGAE-BASED REFUGIUM BIOFILTER

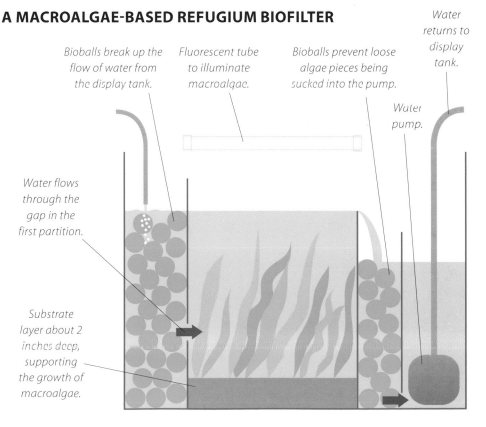

Bioballs break up the flow of water from the display tank.

Fluorescent tube to illuminate macroalgae.

Bioballs prevent loose algae pieces being sucked into the pump.

Water returns to display tank.

Water pump.

Water flows through the gap in the first partition.

Substrate layer about 2 inches deep, supporting the growth of macroalgae.

Above: *Planktivores, such as green* Chromis, *benefit from the addition of a refugium to provide planktonic food.*

(and it can be useful if you want to maximize swimming space in the aquarium), but the living rock remains a valuable adjunct to the algae as a biological filter—as well as bringing all of its other benefits. The living rock is particularly important in the early stages of the aquarium's life, because algae filters take time to become established.

To function effectively as biological filtration, a good growth of algae is required, at least if the aquarium is to house a reasonable population of fishes. This takes time, and the fish population needs to be built up more slowly with this type of system than in an aquarium equipped with a powerful skimmer (which, of course, will run at 100%

efficiency as soon as it is installed). The living rock in the aquarium can help to carry the biological load as the algae is becoming established.

Algae filters—considerations

Algae filters are not without their problems, although these can usually be avoided. To function effectively as a biological filter, fast-growing species of macroalgae are required, and the usual choice is a species of *Caulerpa*. These certainly grow rapidly, but have the disadvantage that sometimes, when their growth is very dense, they can enter a reproductive cycle, in which almost the entire algal mass will disintegrate as they convert their vegetative growth into sexual cells. When this happens, the aquarium can look as if it has been filled with milk, such are the quantities of these cells

that are produced. Although this clears eventually, usually without doing any long-term damage to aquarium inhabitants, the filtration capacity of the algae is lost completely until regrowth occurs.

It appears that these reproductive cycles can be prevented by keeping the algae under continuous lighting and by cropping the algae regularly. Alternatively, different species of algae (that do not undergo the same type of reproductive cycles as *Caulerpa*) can be used—the main requirement is that they are fast-growing.

The refugium as a food source

Even if the aquarium does not use algae for biological filtration, a refugium, whether deliberately planted with algae or not, can be a very valuable addition to the aquarium. While anywhere that fishes cannot reach can function as a refugium, it is best to set up a separate chamber, connected either to the display tank or the sump. Putting a sand and/or gravel bed into the refugium, together with some living rock, will provide the right environment for microfauna to multiply. If the refugium is exposed to light, a variety of algae will grow and you can add some macroalgae (which does not need to be a fast-growing type if it is not being used for filtration). Any type of algae growing in the refugium will increase its productivity.

A productive refugium will provide the aquarium with a supply of living food, primarily composed of small planktonic creatures and slightly larger substrate-dwelling creatures such as amphipods, as well as algal spores (which are essentially a form of phytoplankton).

The tank benefits in a number of ways from the steady supply of planktonic food from the refugium, which is in short supply in many reef aquariums. The continuous drip-feed of plankton from a refugium is ideal for many planktivorous animals, both fishes (such as *Chromis*, *Genicanthus* angels and fairy and flasher wrasses) and invertebrates (particularly corals, which in the wild eat a great deal of plankton), as it mirrors the way their food arrives in the wild.

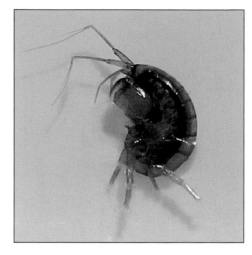

Above: *A refugium can produce huge numbers of small crustaceans, such as amphipods, as shown here.*

A REFUGIUM FOR FOOD PRODUCTION

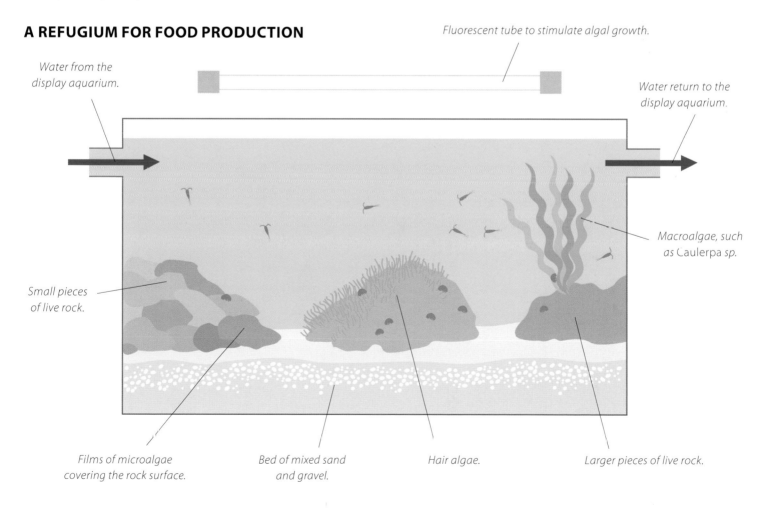

Fluorescent tube to stimulate algal growth.

Water from the display aquarium.

Water return to the display aquarium.

Macroalgae, such as Caulerpa sp.

Small pieces of live rock.

Films of microalgae covering the rock surface.

Bed of mixed sand and gravel.

Hair algae.

Larger pieces of live rock.

Chemical filtration

Protein skimming and biological filtration can deal with most, but not all, waste products produced in the aquarium. Some organic compounds escape these processes and accumulate over time, eventually making the aquarium water appear yellow.

This yellowing not only looks unattractive, but also changes the spectrum and decreases the intensity of light reaching corals. Many dissolved organic compounds are acidic and tend to reduce the pH of the water. They may act as algae nutrients and some may be toxic to invertebrates. These compounds must therefore be removed and this is the role of chemical filtration.

Using carbon

Chemical filtration is usually achieved using activated carbon, which can adsorb a wide range of different substances. There are several methods for using carbon. It can simply be placed in a porous bag somewhere in the system so that water flows around it, without being pumped through it (the "passive" method) or packed into a filter or reactor such that water from the tank is forced through it.

Pumping water through carbon provides very efficient chemical filtration. However, if you use this method, make sure it is in place from the time the aquarium is set up, rather than adding it to an established system. This will avoid rapid reductions in the level of dissolved organics. Such a swift change could lead to a sudden increase in light intensity and shift in spectrum, which could be damaging to corals. "Passive" use of carbon avoids this problem to some degree by being less efficient, so changes occur more slowly.

Another advantage claimed for passive use of carbon is that it is associated with less depletion of trace elements than the pump-through systems. Whether trace element depletion by carbon actually causes

Above: *Activated carbon can be used in a media bag, simply placed in the sump or tank, to provide "passive" chemical filtration.*

Above: *Activated carbon is the most commonly used chemical filtration medium in reef aquariums.*

significant problems is debatable, but as a precaution, some fishkeepers use carbon intermittently (a few days each month, for example) rather than continuously. In practice, all methods—passive and pumped, continuous and intermittent—seem to work, so the choice is a matter of personal preference.

Carbon needs to be replaced periodically, as eventually it becomes saturated with organics and can leach them back into the water. How frequently it needs replacing varies between aquariums and brands of carbon. However, you should only change part of the carbon at any one time to avoid large variations in the levels of organics and trace elements in the aquarium. Changing one third of the carbon (on a first-in, first-out basis) every three months usually works.

Right: *Internal power filters, in the tank or the sump, can be used to pump water through chemical filter media.*

In this filter, granular active carbon or other chemical media can be placed in a special compartment.

Right: *Synthetic chemical filtration media (left) and carbon-impregnated foam pads (right) must have water pumped through them to work effectively—they cannot be used in "passive" mode.*

To facilitate this, you can keep the carbon in three separate media bags, irrespective of whether it is used in passive mode or in a filter. This allows you to replace the media in one bag each time.

Synthetic alternatives to carbon

Carbon is not the only chemical filtration medium available. A number of synthetic alternatives, in the form of foam pads or resin beads, offer similar broad-spectrum adsorption of organic compounds. These media have some advantages over carbon; when they are saturated, no leaching of organics back into the system occurs, and in some cases it is possible to see when they need replacing. Most of these media must have water pumped through them.

The quantities of chemical filtration media required vary widely (even between different types of carbon), so it is best to follow the manufacturer's guidance on how much to use.

Other chemical media

Media that specifically bind phosphates are also sometimes used in reef aquariums. They can be useful in helping to control algae problems. As with other chemical media, these should either be present in the system from day one or be added gradually. Although excessive phosphate levels are very harmful to corals and stimulate blooms of nuisance algae, sudden falls in phosphate concentration can be equally harmful.

Left: *A synthetic resin chemical filtration medium, in the form of beads.*

Maintaining calcium and alkalinity

Sea water is much more complex than a simple sodium chloride solution. It includes over 40 chemical elements, but most are only there in very small concentrations—the so-called trace elements (see page 20). Other components are present at higher concentrations and two—calcium and carbonates—are so important in the reef aquarium that they need to be considered at the design stage.

Corals, calcium, and carbonates

Coral reefs are made of a type of limestone called aragonite—chemically, this is calcium carbonate. Corals and certain algae combine calcium with carbonates, obtained from the water, in their skeletons. Over time, they build the very structure of the reef. In the reef aquarium, abundant supplies of calcium and carbonates are required for

Below: A reef flat such as this, densely packed with fast-growing stony corals, deposits huge quantities of calcium carbonate from the surrounding sea water.

good growth of corals, shelled mollusks, and coralline algae. As these grow they deplete calcium and carbonates from the aquarium water, so aquarists need to be able to replenish the supply.

Apart from providing material to build skeletons and shells, carbonates have another role to play in sea water. Together with related bicarbonate ions, they provide most of the buffer system of sea water, which keeps the pH of the water within a narrow range. This can be measured as the alkalinity of the water—its ability to resist decreases in pH (acidification). As many of the metabolic processes of animals in the aquarium, and biological filtration by bacteria, tend to decrease pH, maintaining the alkalinity of the aquarium water is very important to keep the pH correct.

Maintaining calcium and carbonate levels

Requirements for calcium and alkalinity supplementation vary tremendously between aquariums and the only way

to discover what is needed is to test the water regularly. Sometimes, regular partial water changes may be enough to maintain calcium and alkalinity levels, but most aquariums need more than this. There are several other ways to add calcium and boost alkalinity.

Kalkwasser

Kalkwasser is German for "lime water," a solution of calcium hydroxide. When added to the tank, it delivers calcium and increases alkalinity. Hydroxide ions react with carbon dioxide dissolved in the water, producing bicarbonates. It has an additional benefit; it can precipitate some forms of phosphate, essentially removing these undesirable compounds from the water.

Kalkwasser is highly alkaline, with a pH of about 12, so must be added very slowly to the aquarium, usually using a dosing pump, to avoid raising the tank pH rapidly. Mixing kalkwasser can be hazardous, too: it is made by adding calcium hydroxide or oxide powder to water. These are very caustic, must be handled with great care, and must be kept well away from children.

Kalkwasser also has a relatively low calcium concentration, so if you need to add a lot of calcium, you need to

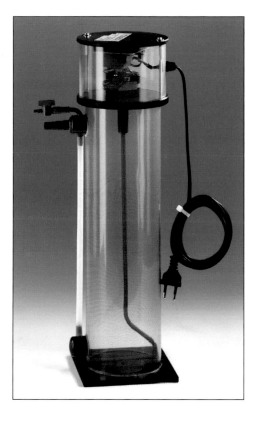

Above: A kalkwasser stirrer provides an automatic way of delivering fresh kalkwasser to the aquarium.

add a lot of kalkwasser. It is usually used to replace water lost by evaporation, and generally the evaporation rate is limited—and so is the amount of kalkwasser that can be added. Despite this, kalkwasser is useful for maintaining calcium and alkalinity (and helping to control phosphate levels), if the tank's calcium demand is not too great.

Calcium chloride, buffers, and two-part supplements

Calcium chloride solutions can deliver far more calcium than kalkwasser, are not caustic, and will not affect the tank pH. Using them, you can add a great deal of calcium into the aquarium in a small volume of liquid. However, adding calcium chloride means adding chloride

ions as well as calcium. This increases the chloride concentration of the water somewhat, although as the main dissolved salt in sea water is sodium chloride, there is already a lot of chloride present and the extra added is small in comparison. However, it does lead to an imbalance between the sodium and chloride ions, especially when a great deal of calcium chloride is added.

Also, calcium chloride does nothing to boost alkalinity, so this needs to be done in another way, usually by adding commercial marine buffers. These are made from sodium carbonate and bicarbonate. When used in combination with calcium chloride, buffers also tend to correct the excess chloride by adding extra sodium.

Two-part, or bi-ionic, liquid calcium supplements are essentially calcium chloride packaged together with a marine buffer, in two separate bottles to prevent calcium carbonate forming and coming out of solution. These provide both calcium and carbonates/bicarbonates, and the chloride from

one component is balanced by the sodium from the other. Solid bi-ionic supplements are also available, which are added to the tank as powder. The components will not react together when dry, so can be stored in this form. The downsides of using these products are their price and the fact that, in addition to boosting calcium and alkalinity levels, they add sodium chloride, which increases the salinity, although this can be corrected by adding fresh water to the tank.

Calcium reactors

The calcium reactor is the technological solution to calcium and alkalinity maintenance. There are many designs, built around the same basic principle. The reactor is supplied with water from the aquarium, carbon dioxide (CO_2) is bubbled through this, which

Below: A section through this aquarium-grown Favia stony coral shows how the calcium carbonate skeleton is fused onto the underlying rock.

Right: *Fast-growing stony corals, such as this* Euphyllia divisa, *need an ample supply of calcium and carbonates in the aquarium water to construct their skeletons.*

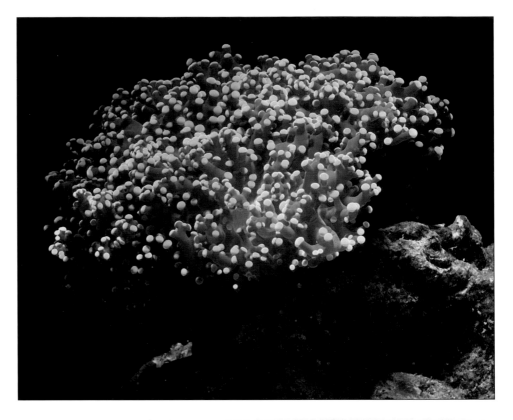

lowers the pH. The water is then passed over calcium carbonate media, which dissolves, and the resulting rich solution of calcium and carbonates is returned to the aquarium.

Calcium reactors can be expensive to buy (particularly as a CO_2 cylinder and regulator valves are also required) and can be tricky to set up and run, often requiring quite a bit of adjustment and maintenance. Despite this, they provide the best way to meet high demands for calcium and alkalinity, for example in tanks with many stony corals. They are reasonably economical to run, requiring only occasional refilling of the CO_2 cylinder and replacement of the media.

Using a calcium reactor

Some care is needed when using calcium reactors. The water returning to the tank, even after passing through the calcium carbonate media, often still contains a significant quantity of CO_2. This has the potential to lower the pH in the aquarium and to stimulate the growth of algae. There are several ways to minimize this. The high-tech solution—effective but expensive and complex to set up—is to use an electronic pH probe linked to a controller that uses a solenoid to stop the gas flow if the tank pH falls below a certain level. Simpler methods include running the effluent from the calcium reactor through a second batch of calcium carbonate media to mop up excess CO_2, and delivering the reactor effluent into a turbulent stream of water so that the gas can be blown off into the air.

HEAVY CALCIUM USERS

Some reef tank inhabitants consume more calcium than others. For example, a 2-inch-long Tridacna derasa *clam has a shell weighing around one ounce. In five years the clam could measure 10 inches long, with a shell weighing two pounds—a single creature requiring the equivalent of over seven ounces of calcium carbonate to be added to the tank per year. This species could grow to at least twice this size and the bigger it is, the more calcium it needs.*

Right: Tridacna derasa

IDEAL ALKALINITY LEVELS

Both calcium and alkalinity are fairly easy to measure, although some test kits express the alkalinity in terms of carbonate hardness (dKH), a parameter borrowed from freshwater chemistry. The alkalinity of water around coral reefs is about 10 milliequivalents (meq)/gallon (carbonate hardness 7–8dKH). In the aquarium, slightly higher levels (up to 18 meq/gallon, or carbonate hardness up to 12dKH) are desirable, and most artificial salt mixes produce water of higher alkalinity than natural sea water.

Left: *Calcium reactors, although complex and expensive, provide a highly effective method of maintaining calcium and carbonate levels, enough to meet the needs of the most demanding aquarium.*

CARBONATE HARDNESS (dKH) TEST

Counting each drop, add KH reagent a drop at a time to a teaspoon sample of tank water. Swirl the tube to mix (1). Initially the sample in the tube turns blue (2). As more reagent is added, the water sample turns yellow (3). Continue counting the drops added until the yellow color is stable. Each drop added from the beginning of the test represents 1°dKH, equivalent to 17.5mg per liter of carbonate.

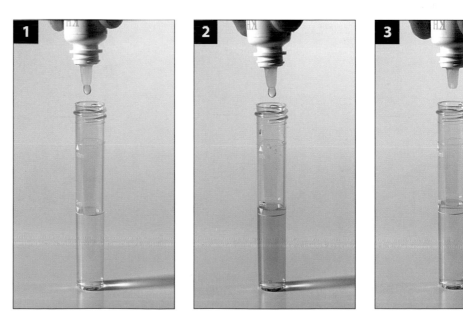

Chapter Three

Invertebrates for the reef aquarium

The reef aquarium system is designed around the needs of the invertebrates, and in particular corals, since these are the most demanding inhabitants of the aquarium in terms of such factors as water movement, light, and water chemistry. Reef aquarium invertebrates more than repay the efforts that go into creating the conditions under which they will thrive and grow. The aquarium owes most of its beauty and interest to the sessile invertebrates: the corals and their relatives in all their diverse colors and beautiful and often bizarre forms, and the spectacular tridacnid "giant" clams, that make up most of the population of a typical reef tank.

In addition to corals, a range of mobile invertebrates can be kept in the aquarium, including crustaceans, worms, and echinoderms, many of which are not only interesting and attractive, but also very useful in the aquarium.

In this section we will look at a range of corals and other invertebrates that are ideal for the reef aquarium, and the specific conditions that each of them requires. This is not a comprehensive list, but aims to provide a selection of creatures that make reliable reef aquarium inhabitants, even in the hands of relatively inexperienced keepers.

Sessile invertebrates are the most important element of the reef aquarium.

Left: *An encrusting* Lobophytum *species, with typical fingerlike projections. In the aquarium, these corals spread quickly across live rock substrates.*

the branches. Another is the cabbage coral (*L. crassum*), which looks fairly similar to *Sinularia dura* but is distinctive among other *Lobophytum* species.

Lobophytum species appreciate moderately strong water movement and bright light, although most tolerate slightly lower light levels well. In common with *Sarcophyton* and *Sinularia*, they produce toxins that can inhibit the growth of other corals. Colonies can grow large (up to three feet across in the wild). Most are relatively soberly colored, in shades of dull pink, gray, cream, brown, or green.

Branching soft corals: *Alcyonium, Cladiella, Capnella, Litophyton,* and *Lemnalia* species

A wide range of branching soft corals are available for the aquarium, and most do well in the reef tank, often growing rapidly, sometimes to large sizes. Most of these corals have relatively subdued colors, in shades of beige, white, and

Above: *Star polyps (Pachyclavularia species) are very attractive, easy-to-keep, encrusting soft corals.*

Right: *Branching soft corals (of various species) can grow rapidly and to large sizes in the aquarium.*

brown. However, their feathery polyps and their movement in the aquarium, swaying in the current (unlike branching stony corals) make them very attractive additions to the tank.

It is difficult to identify many of these corals with any certainty and many are sold under a range of overlapping common names such as cauliflower coral, colt coral, tree coral, and so on. There are some more distinctive species, such as the Kenya tree coral *(Capnella imbricata)*. These corals prefer quite strong currents, but will do well under moderate lighting.

Star polyps:
Pachyclavularia species

Star polyps have a fine combination of resilience, rapid growth, interesting form, and vivid color. The colonies consist of small polyps, typically about .5 inch in diameter, that open from a rubbery encrusting mat (the stolon) that encrusts hard substrates. The stolon is usually purple or maroon. The polyps have eight slender, smooth, or slightly toothed tentacles, often bright green in color, and may have a prominent oral disc that is lighter than the rest of the coral (giving the colonies the look of a mass of twinkling stars). Not all of these species have the prominent oral disc, but even those without this feature are attractive corals.

Star polyps are very undemanding in the aquarium. They grow well under intense illumination or in relatively low light, in strong currents and in quiet water. They even do well in less-than-perfect water quality, although because they are low-growing they can be easily overwhelmed by algae. Star polyp colonies spread rapidly onto neighboring rocks or even aquarium

Left: Clavularia *polyps can spread quickly to colonize neighboring rocks, creating extensive colonies.*

glass under good conditions. Star polyps are not aggressive corals and do not sting their neighbors. They may be damaged by other species, particularly stony corals with powerful stings.

Pachyclavularia can be easily confused with encrusting forms of gorgonians called *Briareum*. These can be distinguished from *Pachyclavularia* by the fact that they tend to form short upright structures from the basal mat, which *Pachyclavularia* do not, and their stolon is not usually so brightly colored. Also, *Briareum* periodically shed sheets of mucus, which is not a habit of *Pachyclavularia*. However, *Briareum* are just as attractive and easy to keep as *Pachyclavularia*, so any misidentification is not a concern.

Anthelia and *Clavularia* species

Anthelia (sometimes called waving hand corals) have polyps with long feathery tentacles at the end of long stalks that rise from an encrusting basal mat. Usually seen in shades of beige or brown, these corals grow vigorously under their preferred conditions of bright light and moderately strong water movement. They are not aggressive and tend to be damaged easily by other corals with stronger stings. *Anthelia* do not appear to feed

on foods offered to them, but may absorb organic substances directly from the water or capture very tiny prey, such as bacteria.

Although not closely related to *Anthelia*, *Clavularia* species (clove polyps) can look quite similar, with encrusting mats from which stems bearing heads of feathery tentacles arise. *Clavularia* tentacles are not usually as elongated as those of *Anthelia* and may be more colorful. Some are green, purple, or yellow, although brown, cream, and grey specimens are more common. *Clavularia* are hardy

Above: Anthelia *are superficially similar to* Clavularia, *but usually have polyps with longer tentacles. They require similar conditions in the aquarium.*

Above: *The bubble coral* (Plerogyra sinuosa) *is usually white or beige. It is best kept in subdued light and gentle currents.*

corals that prefer strong currents and moderate to bright light. Unlike *Anthelia*, *Clavularia* are not easily damaged by other corals, although they are not themselves aggressive.

Stony corals

Stony corals (also sometimes called hard corals) are distinguished from soft corals by their solid limestone skeletons. Soft corals may have small calcareous structures within their tissues, but with a few exceptions do not create skeletons anything like those of stony corals. The need to build such calcareous skeletons indicates one of the key requirements of stony corals in the aquarium: high calcium and carbonate (i.e., alkalinity) levels. Meeting this need can be a

particular challenge in aquariums that have large populations of fast-growing stony corals, which often require calcium reactors to meet their high calcium demand.

Bubble corals: *Plerogyra* and *Physogyra* species

The bubble coral (*Plerogyra sinuosa*) is a unique coral, consisting of large polyps covered, during the day, in large water-filled vesicles—the "bubbles" that give the coral its common name. The "bubbles" increase the coral's surface area tremendously, allowing more efficient photosynthesis. At night, the vesicles contract and are replaced by a mass of stinging tentacles that serve to capture planktonic prey.

Bubble corals are most commonly white or beige, but green, pink, and (rarely) blue-green or blue specimens are seen. In the wild, they are found

in shaded locations, often on vertical surfaces. In the aquarium, they prefer gentle currents and relatively subdued

Left: *A rare, beautiful blue morph of bubble coral (Plerogyra sinuosa).*

Below: *A pearl bubble coral in plankton-feeding mode, with bubbles deflated and stinging tentacles extended.*

macroalgae, and a mass of tentacles arising from the large polyps.

In the aquarium, *Cataliaphyllia* should be positioned, as in the wild, with its skeleton buried in soft, fine sand; placing these corals on rock structures may lead to their tissues being damaged. Filamentous algae need to be controlled, as they can irritate these corals. Gentle currents are preferable, as is moderate lighting. Feeding is beneficial. This species has a powerful sting, and needs to be kept well away from other corals. It grows to around 12 inches across. It is an expensive species, and one to buy only if it can be kept under the correct conditions.

Euphyllia species

Euphyllia species are spectacular stony corals, with large polyps usually covered in masses of tentacles. They have been reef aquarium favorites for many years, although they do need a degree of special care. Most *Euphyllia* species grow large (colonies measuring over 3 feet across occur in the wild), can grow rapidly, and are aggressive species with strong stings and long sweeper tentacles. As a result, they need plenty of space in the aquarium if they are not

lighting, although they can adapt to brighter illumination. They also benefit from feeding with meaty foods. Their tentacles pack a powerful sting, so bubble corals need plenty of space between them and their neighbors. They are usually slow-growing in the aquarium, but are generally hardy, robust species.

Pearl bubble coral *(Physogyra lichtensteini)* is similar to *Plerogyra sinuosa*, but the "bubbles" are much smaller and more numerous. Aquarium care and behavior are much the same as for *Plerogyra sinuosa*.

Elegance coral: *Catalaphyllia jardinei*

This is a unique species that requires some specific care, but can make a magnificent centerpiece for an aquarium. These corals are often extremely vivid, in shades of fluorescent

green, purple, turquoise, or pink, although more soberly colored examples in brown are also found. The tentacle tips are often in contrasting colors to the rest of the coral. In the wild, these corals are usually found in lagoons, with their small, cone-shaped skeletons buried in sand or mud, often among beds of seagrasses or

Right: *Elegance coral (Catalaphyllia jardinei) is one of the most spectacular stony corals, commanding a high price and needing careful husbandry.*

to cause damage to other corals. They make impressive centerpieces where they have adequate space.

Most *Euphyllia* species are light brown or green, often with translucent tentacles, sometimes with contrasting tips. The species divide into two groups: those with solid skeletons and polyps emerging from the entire upper surface, such as *E. ancora* and *E. divisa*; and those with branching skeletons, where the polyps open from the branch tips, for example *E. glabrescens, E. parancora,* and *E. paradivisa*.

Euphyllia species need good water quality, relatively gentle water currents, and moderate lighting. They appreciate feeding with frozen crustaceans (even if this is only catching leftovers from fish feeds). Some species may be adopted as surrogate anemones by clownfishes.

Below: Euphyllia ancora *can grow large and needs plenty of space. Clownfishes sometimes use it as a surrogate anemone.*

This seems to do no harm, at least with small species, such as common clowns and percula clowns.

Some *Euphyllia* species are more robust than others. Hammer, or anchor, coral (*E. ancora*), so called because of its hammer- or anchor-shaped tentacle tips, is one of the more delicate species (it is prone to "brown jelly" infections), whereas frogspawn coral (*E. divisa*) and torch coral (*E. glabrescens*) tend to be hardier.

Trumpet coral: *Caulastrea furcata*

Caulastrea are very attractive, hardy stony corals, found as flat or dome-shaped colonies, with dozens of large polyps. When fully expanded the polyps completely hide the skeleton, which is branched rather than solid, each polyp being at a branch tip. During the day, the polyps inflate to expose maximum surface area to the light, but after dark they deflate and produce long stinging

Above: Caulastrea *species are among the most straightforward of stony corals to keep and are also very easy to propagate.*

tentacles to catch planktonic prey.

There are at least three species of *Caulastrea* (*C. curvata, C. furcata,* and *C. echinulata*) and probably more. *C. furcata* is the most commonly imported species. *Caulastrea* are found in a range of colors and patterns: solid neon green; pale blue-green; brown; brown with green centers, among others. The polyps sometimes develop white radial lines, especially when the coral is kept in bright light.

Caulastrea enjoy bright, but not highly intense, lighting and moderate to strong water flow. *Caulastrea* seem to tolerate less-than-pristine water very well. Under good conditions, *Caulastrea* will grow quickly.

Brain corals: *Trachyphyllia geoffroyi* and *T. radiata*

The two *Trachyphyllia* species are very attractive corals, often highly colorful in shades of neon green (which glow spectacularly under actinic lighting) and sometimes red, although more drably colored specimens are sometimes seen. Both species are hardy, long-lived

Below: A very attractive red morph of Trachyphyllia radiata *shows the characteristic thick, fleshy tissue of this genus of corals.*

Right: A vivid green Trachyphyllia geoffroyi. *Note that these corals are vulnerable to being eaten by fishes that are safe with most other corals.*

Below: When seen side by side, the difference in form between Trachyphyllia geoffroyi *(left) and* T. radiata *(right) are obvious.*

reef aquarium inhabitants. They do not require very high light levels and are relatively tolerant of less-than-perfect water quality. *T. geoffroyi* typically has a cone-shaped skeleton that sits in a sand or mud substrate, and flattish if convoluted soft tissues, whereas *T. radiata* is found attached to hard substrates, has a skeleton with a flat base and has a convoluted, brain-like dome shape. Both species inflate their tissues with water tremendously during the day, whereas at night they deflate somewhat and short feeding tentacles appear. The main threat to these species in the aquarium is predation by fishes; any fish that has even the slightest interest in nibbling corals will usually go for a *Trachyphyllia* as its first choice.

Tankmates for *Trachyphyllia* therefore need to be chosen with great care.

Doughnut coral: *Cynarina lacrymalis*

Cynarina lacrymalis is a stony coral consisting of a single large polyp, usually more or less circular, that has a

relatively small skeleton (perhaps 2.8 inches in diameter). However, it can expand its soft tissues tremendously up to 12 inches in diameter. The skeleton has prominent ridges radiating out from the center, which can usually be seen through the translucent soft tissues, which form prominent swollen

Right: Cynarina lacrymalis, *showing the classic circular form and ridged skeleton. Given the right conditions, it does well in the reef aquarium.*

vesicles above these ridges. The range of colors in *Cynarina lacrymalis* is very wide: browns, reds, pinks, and greens, sometimes with mottling or a polyp center of a contrasting hue.

Cynarina lacrymalis is very hardy in the aquarium, provided that its needs are met. It likes gentle water movement and moderate lighting, and to be fed at night when its tentacles emerge. It should be placed on the sand bed with the polyp facing upwards. It is not an aggressive species, and is easily damaged by corals with stronger stings.

Doughnut coral: *Scolymia* species

The various *Scolymia* species are similar to *Cynarina*, but the ridges on the skeleton are much less pronounced and the tissues expand less dramatically and are not translucent. *Scolymia* are typically mottled in shades of brown, dark green, or dull red. Their aquarium care is similar to that of *Cynarina*.

Brain coral: *Symphyllia* species

The various *Symphyllia* species are hardy corals that can grow large (several yards across in the wild), with heavy skeletons and fleshy polyps. They usually have a brain-like appearance, with sinuous meandering valleys and ridges on a skeleton that is typically shaped like a flattened dome. The ridges between valleys in *Symphyllia* have a groove in their tops, and this feature is usually visible even when the coral is expanded. They are usually muted in color, typically shades of brown, green, and beige. They require moderate water currents, bright light, and plenty of food; they respond rapidly, producing feeding tentacles, when fish are fed with frozen crustaceans. They are not particularly aggressive.

Open brain coral: *Lobophyllia* species

Lobophyllia species are fleshy corals with large polyps that generally have a lobed form. The skeleton may be solid, or the polyps may be carried at the ends of

Below: Although it is not obvious at first glance, this Lobophyllia *has a branched skeleton, with each of its large fleshy polyps at the end of a branch.*

Above: Fungia are free-living corals that can move around on sand substrates. The skeletal structure is usually easily visible.

branches (the branched structure being completely hidden when the polyps are expanded). *Lobophyllia* are typically found in shades of dark red, green, brown, beige, and grey. They are hardy in the aquarium, preferring bright light and gentle water currents. They feed avidly on frozen crustaceans. They are not aggressive and can be damaged by corals with stronger stinging abilities.

Plate coral: *Fungia* and *Heliofungia* species

Fungia and *Heliofungia* are unusual corals. Not only are they not attached to the substrate (they live on sand and gravel beds), but they can actually move around on it, *Fungia* being more mobile than *Heliofungia*. Both have flat or sometimes domed, usually roughly circular skeletons that grow to around 8 inches in diameter, with blade-like septa (ridges) radiating from the central mouth. In *Fungia* the skeleton is usually visible, with the short tentacles appearing between the septa. In *Heliofungia* the tentacles are longer, usually hiding the skeleton. Both are

Above: Acropora *are major contributors to the construction of coral reefs, as seen from these dense thickets of staghorn-like species.*

colorful corals, particularly *Fungia*: colors include green, purple, red, pink, and brown. The tentacle tips in *Heliofungia* are often highlighted in a contrasting color to the rest of the polyp.

In the aquarium, these corals need bright light, gentle water movement, and an open expanse of sand or gravel on which they can be placed (and can wander around). Feeding with frozen fish and crustaceans is also important. *Fungia* are generally hardier than *Heliofungia*, but they also have a more powerful sting and can damage other corals that they come into contact with, except for other *Fungia*, which seem to be immune.

Acropora species

Acropora is probably the single most important genus of corals with respect to building reefs. There appear to be

dozens of species, although it is hard to say exactly how many, as *Acropora* can change their growth forms under different conditions. *Acropora* are found in many colors, including purple, blue, pink, green, brown, yellow, and magenta, and many colonies are multicolored, often having contrasting branch tips (each of which has a polyp at the end, a distinguishing feature of almost all *Acropora* species). They will often change color when introduced to the aquarium, sometimes becoming more colorful, sometimes less. Colonies

Right: *Some* Acropora *species form spectacular table-like structures in the wild—and, if conditions are right, in the aquarium.*

may occur in several forms: staghorn, clustered branches, table-like, bushy and bottlebrush-like, among several others.

In the aquarium, *Acropora* need very strong water flow, high levels of calcium and carbonates, and very intense lighting. They can be very difficult to keep, particularly wild-collected colonies (which are most likely to be lost shortly after importation). They can quite often fall victim to problems such as tissue recession and rapid tissue necrosis if conditions are less than perfect (and sometimes even when conditions apparently are perfect!).

Fortunately, *Acropora* can be grown fairly easily from small fragments of colonies and they have been the subjects of mariculture in the tropics and aquarium propagation, both on a small commercial scale and for exchange of corals between enthusiasts. Maricultured colonies seem to be easier to keep than wild-collected ones, and aquarium–propagated *Acropora* are even more likely to succeed in the reef tank. Some types of *Acropora* have been aquarium propagated over several generations (fragments taken from colonies grown from fragments and so on). These are generally quite robust

Above: *A beautiful purple plating form of* Montipora. Montipora *are also found as encrusting and branching colonies, in a wide range of colors.*

in the aquarium, tolerating less than perfect conditions, although they will show much better color and form when lighting, water flow, and water quality are ideal.

Acropora grow fast under favorable conditions (several inches per branch per year is possible) and may quickly reach a point where they require pruning, as the colonies grow into other corals, reach the water surface, or begin to reduce the water flow in the aquarium by blocking currents.

Montipora species

The various species of Montipora cover many different growth forms, from branching, encrusting, plate-like, and vase-shaped to whorls. The color range of Montipora is also very wide, including various shades of brown, green, orange, purple, maroon, pink, and blue. Areas of rapid growth (branch tips and the edges of plates) are usually lighter-hued or may be in different colors

to the rest of the colony. Montipora polyps are very small. On branching and encrusting forms they are densely packed, giving the colonies a furry or velvety appearance. On plating forms, the polyps are more widely spaced or may not be visible at all.

Montipora have light, porous, brittle skeletons that are easily broken, although the resulting fragments will readily grow into new colonies. Most Montipora species are very adaptable and although very intense light and

strong water movement tend to favor the growth of the most colorful colonies, most species will grow under lower light levels. Under good conditions growth can be very rapid, with sizeable colonies growing from small fragments in only a few months.

Montipora are not aggressive corals, despite their rapid growth. They almost invariably lose territorial conflicts with other corals. They are generally very hardy in the aquarium.

Porites species

Best known for forming very large, boulder-like colonies on the reef, the various species of Porites make excellent, hardy aquarium inhabitants, provided that their needs for intense light and strong water movement are met. Porites are found in a number of forms, including branching and encrusting colonies, as well as the

Below: *This is a branching form of* Porites; *yellow is one of the more common colors in these corals.*

Above: Pocillopora damicornis *is hardy and grows fast in the aquarium, provided it has enough light and water flow.*

more familiar boulder-like coral heads, and in a range of colors, notably brown, purple, green, blue, and yellow. They superficially resemble *Montipora*, but their skeletons are much stronger and denser. *Porites* are unusual for stony corals in that they periodically clear the colony surface by shedding a skinlike layer of mucus, in a similar way to some soft corals.

In the aquarium trade, *Porites* are often encountered in the form of brown colonies studded with small, brightly colored fanworms, known as Christmas tree worms (*Spirobranchus giganteus*). These combined colonies make very attractive aquarium inhabitants. As long as the coral is healthy, the worms will generally survive (they appear to get at least some of their nutrition from the coral), provided they are not eaten by fishes; *Centropyge* angels are often the culprits. Other *Porites* forms, usually branched colonies, are also available.

Porites grow quickly under good conditions (this is most obvious in the branching forms), but do not seem to be aggressive toward other colonies.

Pocillopora damicornis

Pocillopora damicornis is a branching stony coral that typically forms rounded bushy colonies with densely packed branches. The polyps are small, with translucent tentacles that give colonies a fuzzy look. Colors are usually brown, green, or pink or combinations of these colors. In the wild, this species grows in a wide range of reef habitats and is very variable in appearance.

Pocillopora damicornis requires strong water currents and high water quality. Ideally, illumination should be very intense; this species will tolerate less bright light, but may not be so colorful under such circumstances. Under good conditions, it can grow very rapidly. It can be quite aggressive, stinging corals that are in its path as it grows, so give it plenty of space. *Pocillopora damicornis* is easily propagated from small fragments.

Bird's nest coral: *Seriatopora* species

Seriatopora are very distinctive stony corals, with thin branches that ramify and entangle to create the characteristic bird's nest appearance. The polyps are small and arranged in lines along the branches. The most common colors are brown and pink, although green colonies are also found. *Seriatopora hystrix* has very thin branches with very

Below: Seriatopora hystrix *is a finely branched, rather fragile coral that is relatively straightforward to keep in the aquarium. Pink forms are common.*

sharp tips. The branches of *Seriatopora caliendrum* are somewhat thicker and the tips are not so sharp.

Strong (but not extremely intense) lighting and reasonably brisk water flow suit *Seriatopora* species best. Water quality needs to be high. Growth can be rapid and broken branches (almost inevitable given the thin, brittle skeleton) readily grow into new colonies. These corals are not particularly aggressive and can be easily damaged by more competitive species.

Club finger coral:
Stylophora pistillata

Stylophora pistillata has thick branches with rounded ends that give it a very distinctive appearance. The polyps are very small, their tentacles making the branches look fuzzy. This species can be very colorful, in shades of brown, green, purple, blue, and pink. The polyps may be a contrasting color to the coral's branches. They can grow rapidly and are easily propagated from fragments. In the aquarium they are quite adaptable to different levels of light and water movement (although the colony color and shape may change over time in response to different conditions), but high water quality is required. This is not an aggressive species.

Turbinaria species

Turbinaria are attractive stony corals, many of which are very straightforward to keep. Typically found in bowl, plate, goblet, or scroll shapes, but occasionally in columnar form, *Turbinaria* range in color from beige through greens and browns to yellow, the latter being particularly common in one species, *T. reniformis*. The different species vary considerably in the size of their polyps.

Turbinaria peltata and *T. patula* have polyps up to 1 inch in diameter, which in some specimens hide the skeleton in a mass of tentacles when expanded, whereas the polyps of *T. reniformis* and *T. frondens* are tiny.

Turbinaria corals can be found in a wide range of reef areas, from turbid lagoons to clear reef slopes. Their great diversity of shapes reflects this. In shallow water, with high light intensity, they may form scrolls or columns, whereas plates or bowls

Above: Stylophora pistillata *is a unique coral. Its typically furry appearance is due to the very fine tentacles of the polyps.*

are more common in forms from deeper water with less intense light. This versatility and their tolerance of different conditions make them hardy reef aquarium inhabitants, particularly *T. peltata* and *T. patula*. All are non-aggressive and grow steadily in the aquarium, reaching impressive sizes in time. Generally, *Turbinaria* like fairly bright light and good water movement. They are reasonably tolerant of less than perfect water quality.

Faviid corals: moons, brains and pineapples

The *Faviidae* family of stony corals includes many excellent, if sometimes difficult to identify, species for the reef aquarium. Most require similar care in the aquarium—moderate to strong water movement and bright light, together with high water quality and an occasional meal of frozen crustaceans, delivered at night, when these corals extend their feeding tentacles. Most faviid corals produce sweeper tentacles with powerful stings and need to have a clear zone around them to prevent other corals being damaged.

Under good conditions, faviids grow steadily, if not rapidly, and will fix themselves down onto hard substrates

as they do so. The colony form in most species is either encrusting or domed. In the wild these corals can grow to several yards across. Faviid corals often bleach when exposed to high water temperatures, but will usually recover in time. They can also suffer from tissue

Below: Turbinaria *are very hardy stony corals that are often found in bowl- or vase-shaped colonies, although they can form plates or scrolls.*

Left: This unusually colored Favia *shows the characteristic arrangement of polyps in this genus. Each of the polyps has its own separate walls.*

Above: Favites *are very similar to* Favia, *except that the polyps share common walls, giving a more honeycomb appearance.*

recession and "brown jelly" infections, but these problems can usually be treated using freshwater dips (see page 196) and prevented by proper placement of the coral with respect to water flow and light.

Favia species, often referred to as pineapple or moon corals, have polyps typically one-half inch or so in diameter. Each has its own outer walls—a feature that distinguishes *Favia* from some other corals in the family. Colors are usually brown and green—often a brown outer part of the polyp with a green center. The green pigments in *Favia*, as in other members of this family, usually fluoresce vividly under actinic-type lighting.

Favites have one key difference to *Favia*, namely that the polyps share their outer walls with adjacent polyps, giving *Favites* more of a honeycomb

appearance than *Favia*. In other respects they are very similar to *Favia*, and tend to be sold under the same common names.

Platygyra, Oulophyllia, and *Goniastrea* are all sold as brain, or sometimes maze, corals. Sometimes they can be hard to distinguish from each other, particularly *Oulophyllia* and *Goniastrea*. They all have polyps that are joined together in meandering valleys, without the cross-walls that separate *Favites* polyps, for example. The result is a brainlike or in more extreme cases (these are usually *Platygyra*) a maze-like appearance. *Platygyra* typically have longer and more meandering valleys, with more polyps, than *Oulophyllia* and *Goniastrea*. *Goniastrea* typically have smaller polyps than *Oulophyllia* and do not have such high walls between the polyps. *Goniastrea* are extremely hardy corals in the aquarium, whereas *Platygyra* are less so, being rather more susceptible to tissue recession and brown jelly infections than other faviids.

BEWARE—CHEMICAL WEAPONS

Some zoanthids, notably Palythoa *and* Protopalythoa, *produce potent neurotoxins called palytoxins, in their mucus, presumably to discourage predators. Palytoxins seem to do no harm to other aquarium inhabitants, but may pose a threat to fishkeepers. It is best to avoid handling zoanthids, but if you do, wash hands thoroughly afterwards (always a good practice after touching corals).*

Colonial anemones

Often thought of as soft corals because of their lack of a calcified skeleton, mushroom anemones and zoanthids (both justifiably very popular in the reef aquarium) are more closely related to stony corals than to the "true" soft corals, such as leather corals. Both belong to the zoological subclass *Hexacorallia* (corals with six, or multiples of six, tentacles per polyp) like the stony corals. True soft corals have eight tentacles per polyp and are classified in the subclass *Octocorallia*.

Mushroom anemones: *Discosoma, Rhodactis, Actinodiscus, Ricordea,* and *Amplexidiscus* species

Although their basic form is always the same (a flattish disc, attached on the underside to the substrate), mushroom anemones have such a wide range of colors, patterns, and textures that probably only *Acropora* stony corals can rival them for variety. They can be brown, blue, turquoise, green, orange, red, beige, or yellowish, or almost any combinations of these. Many have contrasting spots, stripes, or even marbling. They may be smooth, patterned with raised dots, or have tentacles ranging from the simple to the furry. They range from 1 to 18 inches in diameter.

The mushroom anemones usually kept in the aquarium are virtually impossible to identify to species level; their taxonomy is controversial even among marine biologists. It currently seems that *Rhodactis, Discosoma,* and the more distinctive *Ricordea* are genuine genera. The "elephant ear" (*Amplexidiscus fenestrafer*) also appears to be a genuine species.

Above: *Mushroom anemones are found in a huge variety of colors, patterns, and textures, and look wonderful when kept together in mixed groups.*

Fortunately, all the commonly kept mushroom anemones require similar aquarium conditions. They prefer quite bright, but not extremely intense lighting, gentle water movement, and reasonably good water quality, although they do not need the perfect water that some other corals require.

Mushroom anemones are among the most resilient of invertebrates in the aquarium. They usually reproduce readily under favorable conditions, generally by dividing or budding off

Three genera of zoanthids are of interest to reef aquarium keepers, although precise identification, as with many other corals, is often very difficult. *Palythoa* are large zoanthids (up to 1 inch diameter), often bright or even metallic green (although they may also be brown or black). The tentacles are short and the polyp bases are embedded in a leathery mat known as a coenenchyme, which is attached to the substrate. *Palythoa* colonies are often smaller than those of some other zoanthids, seldom measuring more than 12 inches across.

Protopalythoa are closely related to *Palythoa* (indeed, some biologists think they should all be included in *Palythoa*), with polyps of similar size and range of colors. However, *Protopalythoa* polyps are separate and not embedded in a coenenchyme and tend to have more tentacles than those of *Palythoa*.

Zoanthus are smaller than most *Palythoa* and *Protopalythoa*, up to

Below: *Zoanthus polyps grow in dense mats, often in bright colors; the colonies have a more compact appearance under intense light and in strong currents.*

new individuals. They are not generally aggressive toward other mushroom anemones, but some of them can inhibit some stony corals. Some of the larger species can also catch small or slow-moving fish, so the large "elephant ear" species should not be kept with small gobies or similar fishes.

Zoanthid or button polyps: *Zoanthus, Palythoa,* and *Protopalythoa* species

Zoanthids are small colonial polyps that resemble miniature sea anemones, ranging from about 1/4 inch to 1 inch in diameter. They can form quite large colonies, sometimes covering several square yards of substrate in the wild.

Most zoanthids thrive in very intense light and vigorously moving water. They can tolerate less-than-perfect water quality and very high temperatures. Most species are relatively inoffensive toward other corals, and indeed are often inhibited by more aggressive species. However, when conditions are favorable, they reproduce rapidly by division and can sometimes overgrow corals that lack strong stings or other defenses.

Zoanthids may be very brightly colored, often in metallic green or turquoise, red, orange, or yellow. Others are beige or brown, but often have contrasting tentacle tips or markings. Large colonies can look spectacular.

.4 inch in diameter, with short, stubby tentacles. They are found in a wide range of colors, often with contrasting shades on the mouth or tentacles, or even with patterns on the oral disc. Some can form very extensive colonies.

Clams

There are nine species of *Tridacna* giant clams, some more giant than others. They range in maximum size from the 6-inch *T. crocea* to the 4-foot *T. gigas*, the true giant clam. They show an interesting pattern of convergent evolution with corals, in that they harbor zooxanthellae in their tissues and so can make use of the abundant sunlight on the reef, rather than relying on filter feeding, such as most other bivalve mollusks.

Not all *Tridacna* species are always (or in some cases ever) available in the aquarium trade, but the four species here are not usually too difficult to obtain. Most clams that enter the aquarium trade today are farmed in the tropics; wild *Tridacna* clams suffered from overfishing (for food) in some

Left: Tridacna derasa *is the second largest clam (after* T. gigas*). It is also probably the best species for the aquarium—if space permits.*

areas, and the clam farming industry evolved in response to this, primarily to feed the continued demand for clams as food in Asia. A useful by-product of the industry is the availability of clams for the reef aquarium.

Clams do well under ordinary reef aquarium conditions, but they do need very intense lighting, preferably from metal-halides. Only *T. derasa* can do well

Below: Tridacna crocea *is probably the most colorful of the giant clams, as well as the smallest. It needs very intense light.*

under fluorescent lighting, and even then a substantial bank of lamps is required. Intense lighting is particularly important because to look their best, clams need to be viewed from slightly above (this shows off the mantle colors to best effect), so need to be placed on or close to the base of the tank. Calcium and alkalinity levels need to be kept high, as clams deposit substantial quantities of calcium carbonate into their shells. Clams also benefit from good algae control, as algae growing on their shells may inhibit them from expanding their mantles properly. Some fish (often algae eaters, such as blennies and *Centropyge* angels) can nibble at the mucus produced by clams and, again, prevent proper expansion.

Tridacna derasa is the second-largest of the family, quickly growing up to 2 feet across. The lighting requirements of this species are less rigorous than for the other tridacnid clams, but it will still grow faster and look more colorful under more intense illumination. *Tridacna derasa* grows well on a sand substrate and (perhaps because of its weight when well-grown) does not attach itself to the substrate. The mantle

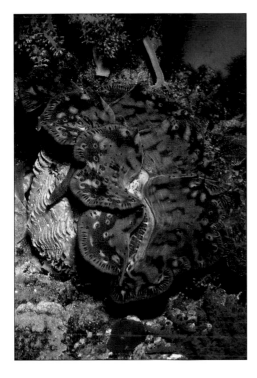

Above: Tridacna maxima *is found in a wide range of different colors and patterns; this attractive metallic coppery gold is unusual, but such specimens turn up occasionally.*

color and pattern in this species are usually quite distinctive (typically dark brown with thin, irregular, light stripes that turn golden under bright light) and the shell is smooth, with no protruding plates, ridges, or scales.

Tridacna squamosa is another clam that can grow quite large, up to 1.3 feet long. It has a distinctive shell, with large, widely spaced scales jutting out from its ribs. The mantle often has a spotted or speckled appearance, although there is considerable variation between individuals. Typical colors are brown and white, although some specimens have a bluish mantle. Tridacna squamosa will often anchor itself to a hard substrate with a weak byssal attachment, but will also live happily on a sand bed. It requires more light than T. derasa, but is

not as dependent on extremely intense lighting as T. maxima and T. crocea.

Tridacna maxima is a little smaller, growing to 12 inches, and has extremely varied colors and patterns on the mantle. It can be highly colorful, in shades of blue, turquoise, green, and gold, although specimens that are predominantly brown or even black can be found. This species prefers to attach itself to hard substrates. In the wild it often embeds itself into dead coral or soft rock. The shell is more elongated than that of T. squamosa, and is covered in dense rows of protruding scales. T. maxima is a little more demanding than T. derasa and T. squamosa in the aquarium, particularly with respect to light. Intense illumination is essential for this species.

Tridacna crocea is the smallest of the "giant" clams, reaching an absolute maximum of 6 inches. It is also the most colorful, with mantles colored in dazzling blue, turquoise, green, and

gold, in a wide range of patterns. The shell is similar to that of T. maxima, covered in rows of thin scales. This species attaches to, and burrows into, hard substrates, often having almost all of the shell burrowed into rock or coral heads. It achieves this both by a grinding action of the shell and secretion of acids that dissolve the substrate. In the aquarium it needs a hard substrate to attach to, which it does by means of strong byssal threads. It is essential not to damage these. Of all the clams, T. crocea is the most dependent on intense lighting and really needs to be kept in a tank with powerful metal-halide lighting.

Below: Tridacna squamosa *usually has a mottled pattern on the mantle, and a ridged shell. It is a good choice for the reef aquarium, growing big enough to make a spectacular specimen, but not so large as to be hard to accommodate.*

Mobile invertebrates

Corals and clams may be the most conspicuous inhabitants of the reef aquarium, but many other invertebrates can also be kept. Unlike corals, clams, and other sessile invertebrates, most are not fixed to the substrate but are free-living or mobile. A great variety of mobile invertebrates will live in a marine aquarium, but not all are suitable for the mixed community of a reef tank. Certain species (many starfishes, for example) may eat other invertebrates; others may inadvertently damage their tankmates (sea urchins, for instance, may puncture corals or knock them off rocks). Some species may eat fishes (the large green brittle star *Ophiarachna incrassata* is a notorious predator of small fishes) or even poison them (some sea cucumbers produce poisonous eggs, which fishes eat with fatal consequences).

Many mobile invertebrates are not just interesting and attractive; they are also useful. They may graze nuisance algae, clear up uneaten food, and help to prevent detritus accumulation. The mobile invertebrates featured here fall into three groups: mollusks, echinoderms, and crustaceans.

All those discussed are sensitive to changes in salinity and so need to be introduced to the aquarium very carefully, using a slow equilibration procedure (see page 158).

Mollusks

A variety of mollusks can be kept in reef aquariums, but algae-eating snails are the most popular and useful choices.

Algae-eating snails

Various snails are commonly kept in the reef aquarium for a strictly functional purpose, namely the control of nuisance algae. However, they are interesting creatures in their own right and deserve more attention than they sometimes receive. These snails are generally hardy and long-lived if conditions are right, which mainly means that there is sufficient algae growing in the aquarium and an absence of predators.

The most common mistake in keeping algae-eating snails is overstocking, generally in an attempt to control or prevent algae problems. The usual outcome is that some of the snails will starve and die, until the population is small enough to be sustained by the available supply of algae.

These snails will not attempt to eat sessile invertebrates, but they can occasionally cause problems by knocking corals off rocks or by breaking branches as they pass by. Most of

Above: Astraea *and* Lithopoma *snails are very similar, and are among the best algae-grazing snails for the reef aquarium, being smaller than many others.*

them are very strong and can push surprisingly heavy objects out of their way as they graze. It is the larger species that are most often associated with such problems.

Astraea and *Lithopoma* species

Astraea and *Lithopoma* (they are difficult to distinguish) are among the best grazing snails for the reef aquarium. They are small (up to about 1⅛ inches in diameter), which enables them to get into tight spots to graze algae and makes them less likely to cause problems by knocking over corals. They seem to be rather hardier than *Turbo* snails, but this may be simply the result of being smaller and less likely to run out of food in the aquarium. Many *Astraea* and *Lithopoma* have triangular projections from the sides of the conical shell, and seen from below these give the shell a starlike shape. A reasonable stocking rate for *Astraea* and *Lithopoma* is one snail per 2.5 gallons of tank capacity.

Turbo species

Many snails (including *Astraea* and *Lithopoma* species) are sold as "turbo snails," but true *Turbo* snails have a coiled, conical turban shape to their shells (hence their scientific name). The shells have a circular aperture from which the snail emerges. The aperture is closed using a calcareous flap called an operculum. The snails are found in most seas (tropical and otherwise), and can grow to around 6 inches in diameter, although those typically kept in the aquarium are sold when they measure around 1⅛ inches in diameter and grow to about 3 inches. They need a plentiful supply of diatoms and other low-growing algae. A suggested stocking level is no more than one *Turbo* snail per 10 gallons of tank capacity.

Trochus species

Trochus species are quite large snails, up to 3 inches in diameter, with regular, conical shells that often have subtle, spiral ridges. These are hardy snails, but their large size means that they are more likely to damage corals by toppling them from rocks or even breaking branches as they pass by. As befits their size, they have hearty appetites, so should not be overstocked: one snail per 26 gallons of tank capacity should be the maximum stocking rate.

Left: Trochus are large snails that need a plentiful supply of algae to keep them well fed; they should not be overstocked.

Tectus species

In common with *Trochus*, *Tectus* species are quite hefty snails, growing to at least 3 inches in diameter. They have smoothly conical shells and are quite active, fast-moving creatures (by snail standards). Corals in tanks with *Tectus* snails need to be well secured on their rocks, as these snails are quite capable of dislodging even quite heavy corals. If they have enough algae to eat they are generally hardy and long-lived. As with *Trochus*, a maximum stocking rate of one snail per 26 gallons of tank capacity is advisable.

Nerita species

Nerite snails have turban-shaped shells that are usually dark brown or black. They are generally small, up to 1½ inches and are good, hardy, reef aquarium grazers. However, some species are found in tidepools in the wild and will sometimes crawl above the water level—or even out of the

Above: Larger snails, such as this Tectus species, are efficient grazers, but can be clumsy and may damage corals.

Left: Nerita snails cleaning the aquarium glass; they are small, robust animals that sometimes climb out of the tank.

tank. Stocking levels for *Nerita* species should be a maximum of one snail per 2.5 gallons of tank capacity.

Echinoderms

Many types of echinoderms, including sea urchins and sea cucumbers, may be kept in reef tanks, but many are difficult to keep or can cause problems for their tankmates. (For example, some sea cucumbers when injured release toxins that can kill fishes.) Brittle stars and serpent stars are probably the best reef tank inhabitants, and a small selection of other starfishes also do well and do not cause problems in the aquarium.

Starfishes

Most starfishes make poor reef aquarium inhabitants. They are either predatory on corals or, at the other extreme, fail to feed and starve to death; the natural diet of many starfishes is poorly understood. The two species described here usually do reasonably well, but both are best kept in larger tanks, preferably mature systems, where they have the best chance of finding the types of food they need.

Blue starfish (*Linckia laevigata*)

This beautiful species is found in various shades of blue. Small individuals, 4 to 5 inches in diameter, are usually offered for sale, but this species grows surprisingly large, up to 16 inches across, and needs to be kept in a large aquarium where it will have a good chance of finding an adequate supply of food. When buying this species, watch out for small parasitic mollusks embedded in the tissues. These are often hard to see, as they may be the same color as the starfish. They can be removed with forceps, but it is better not to have to do this.

Spotted starfish (*Linckia multifora*)

This small starfish grows to 4 inches across and is attractively colored in mottled red, white, orange, or blue. Perhaps as a result of its relatively small size it usually does well in the reef aquarium; it can presumably find enough food to sustain itself. Even so, it is best kept in larger tanks where it is least likely to run out of food. It appears to be harmless to other reef aquarium inhabitants.

Brittle stars and serpent stars

Serpent and brittle stars are valuable additions to a reef aquarium, as they are excellent scavengers that clear up excess food and any dead animals. There is no real scientific distinction between serpent and brittle stars; a strictly non-scientific way of telling them apart is that serpent stars have smooth arms and brittle stars have spiny arms. Most species grow to about 8 to 12 inches in diameter, but some can become much larger. They are generally hardy and long-lived. Although many are nocturnal, most will emerge from their hiding places at feeding time. In time many, particularly serpent stars, will become more outgoing during the day.

Most brittle stars are black, brown, or gray, but they are still interesting and useful animals. Serpent stars are found in a wider range of colors. Green-and-white, brown-and-white, reddish brown, green, and black are common.

There is one serpent star to avoid: *Ophiarachna incrassata* is a green species that grows to at least 19 inches

Right: The spotted starfish is probably the best species for the typical reef aquarium. Damaged limbs (the lower right arm in this one) usually regenerate without problems.

Left: The beautiful blue starfish (Linckia laevigata) should be kept in a large aquarium so that it can find enough food—it may starve in smaller tanks.

Left: Brittle stars are seldom seen in the reef aquarium during daylight hours, but are extremely useful as general scavengers of uneaten food and any dead animals.

across and eats fishes. It has thicker arms than most others, and the arms have short, blunt, yellow-and-black spines along them.

Black brittle star (*Ophiomastix variabilis*)

This Pacific brittle star is very attractive, with velvety black arms with short spines and bright yellow markings. It is usually hidden during the day, but remains a useful scavenger of uneaten food and dead animals, while not being a threat to its tankmates. It grows to around 8 inches.

Spiny black brittle star (*Ophiocoma* species)

Ophiocoma come from the Caribbean, and are usually black or dark brown. Their relatively short arms are covered in long, blunt spines. They grow to around 8 inches diameter. In the aquarium they tend to be nocturnal, but are still very useful as scavengers. They are usually harmless to their tankmates.

Serpent stars (*Ophiolepis* and *Ophioderma* species)

Ophiolepis and *Ophioderma* species include most of the serpent stars commonly imported for the aquarium. They range in color from solid deep reds and browns, to greens and combinations of these colors. Many are white, with green or deep red bands on the arms and markings on the central disc. All are excellent aquarium inhabitants and usually become quite extroverted—at least at feeding time. They eat most foods but do not pose a threat to fishes or other invertebrates. Most grow to 6 to 12 inches across.

Red serpent star (*Ophioderma squamosissimus*)

This species comes from the Caribbean and is not only more colorful than most other serpent stars, but also larger, growing to around 16 inches across. Once established in the aquarium, it is hardy and long-lived, and seems to be harmless to other aquarium inhabitants. In the aquarium it often becomes quite outgoing, resting on the substrate in the open even during the day.

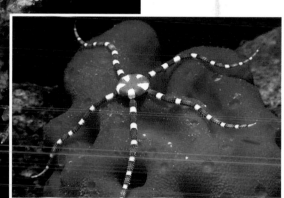

Above: Serpent stars are found in a wide range of colors and patterns and make attractive, long-lived reef tank inhabitants.

Left: The beautiful red serpent star is very hardy once established in the aquarium, and is seen more often during the day than other serpent and brittle stars.

Above: *The Pacific form of cleaner shrimp* (Lysmata amboinensis) *is an attractive addition to the reef aquarium; it is more extroverted than most other* Lysmata *shrimps.*

Crustaceans

The range of crustaceans that could be kept in a marine aquarium is very wide, but relatively few are ideal for a reef aquarium housing the usual mix of corals and fishes. Most true crabs and many shrimps, as well as large hermit crabs, prove to be vandals in the reef tank, damaging or eating corals and sometimes even predating on fishes. However, the selection described here make generally good reef aquarium inhabitants. One cautionary note with crustaceans, particularly shrimps, is that they tend not to tolerate high water temperatures.

Cleaner shrimps (*Lysmata amboinensis* and *L. grabhami*)

Many species of shrimps will, at least occasionally, clean fish of parasites, including most of the other shrimps described here. However, two species are usually sold specifically as cleaner shrimps. These are *Lysmata amboinensis* from the Pacific and *L. grabhami* from the Caribbean. The two species look very similar. They are very attractive scarlet shrimps, with a bright white stripe along the back, and white legs and antennae. Both species grow to about 2 to 3 inches long. It is possible to differentiate between them by looking at the markings on the tail. In *L. grabhami*, the white stripe along the back extends to the end of the tail, and the outer edges of the tail also have a thin white margin, whereas in

L. amboinensis the white stripe on the back stops at the base of the tail fan, which has a wedge-shaped white mark on it, and white spots at the front and rear of each side.

Cleaner shrimps are perhaps the most extroverted of the shrimps that are usually kept in the aquarium, spending much of their time out in the open. They are sociable creatures and can be kept in groups. Their cleaning behavior is often maintained in the aquarium, and it is fascinating to watch them grooming their fish "clients." In addition to the food they obtain through cleaning, they will eat just about any kind of fish food and act as general aquarium scavengers. They will sometimes steal food from coral polyps.

Female cleaner shrimps quite often produce clutches of bright green eggs,

CLEANERS

Lacking arms and legs, and in most cases not being flexible enough to be able to bite external parasites off their bodies, fishes rely on a variety of other creatures to remove these organisms. A range of fishes, including wrasses, gobies, and juvenile angelfishes and even some butterfly fishes, will provide this service, as will many shrimps. As well as removing parasites, many cleaners also remove loose scales and damaged skin. Cleaners generally enjoy protection from predation (many of their clients could easily eat them) to the point where they may be allowed inside larger fishes' mouths to clean their teeth. Most cleaners advertise their services by distinctive markings. Many cleaner shrimps (such as Lysmata debelius, L. amboinensis, L. grabhami, *and various* Stenopus *species) have very bright white antennae. Shrimps often reproduce their cleaning behavior in the aquarium, which is fascinating to observe, as well as presumably being beneficial for the fishes involved.*

which they carry on the underside of the abdomen. The larvae provide planktonic food for the aquarium.

Peppermint shrimp (*Lysmata wurdemanni*)

The peppermint shrimp is a rather secretive species; once introduced, it may seldom be seen again, at least during the day. However, this is not a problem, as peppermint shrimps are not usually kept for their appearance, but because they are useful. They eat *Aiptasia* anemones, even those that are much larger than the shrimps themselves. Peppermint shrimps are probably the reef aquarium's best biological control for *Aiptasia*.

The peppermint shrimp has a transparent grayish body overlaid with criss-crossing red lines. Like other *Lysmata* species, it is straightforward to keep, eating most foods.

This species is not territorial and will live in a group with its own kind, and share its tank with other types of shrimp. In addition to their role in controlling *Aiptasia*, peppermint shrimps are good scavengers of uneaten food and other edible debris. They spawn readily in the aquarium,

Below: *The peppermint shrimp is one of the most useful shrimps to keep in the aquarium, thanks to its predation of* Aiptasia, *but it may never be seen during the day.*

releasing clutches of planktonic larvae that provide live food for fish and corals.

Peppermint shrimps require special care when first added to the aquarium. They are even more sensitive than other shrimps to changes in salinity and need a very slow equilibration between shipping water and tank water when they are introduced to the aquarium. Taking two to three hours over the acclimatization process (even slower than the usual slow method used for shrimps) is wise. Once established in the tank, peppermint shrimps are reasonably hardy, but are particularly easily affected by high water temperatures.

Fire shrimp (*Lysmata debelius*)

Also known as the blood shrimp, *Lysmata debelius* from the Indo-Pacific Ocean is one of the most beautiful shrimps, a solid deep red with a few white spots, white antennae, and white tips to its three pairs of walking legs. It grows to around 1 inch. It acts as a cleaner shrimp in the wild, and sometimes cleans fishes in the aquarium.

Unfortunately, it is a rather shy species under most aquarium conditions, preferring dimly lit areas (it is a cave-dweller in the wild). It seldom ventures out during the day in brightly illuminated reef tanks, although it will sometimes dash out from its lair at feeding time. However, if part of the aquarium is given over to a large cave, the fire shrimp may be more visible.

Left: The fire shrimp is one of the most beautiful crustaceans for the aquarium, but it is shy. A large cave will make it feel at home—and keep it out on show.

In common with other *Lysmata* shrimps it can be kept in groups of its own kind, and with other species of shrimps. It scavenges a variety of aquarium foods.

Boxer shrimp (*Stenopus hispidus*)

Boxer shrimps (*Stenopus hispidus*) grow to about 4 inches long and a similar size across their large claws. They are quite spectacular creatures, their white body and limbs having red banding. Females have blue-purple areas (the ovaries) visible inside the body through the shell. They act as cleaners for large fishes in the wild, but this behavior is seldom observed in the aquarium.

Boxer shrimps are among the hardiest of shrimps in the aquarium. However, they are territorial and should be kept either singly or in a mated pair (the latter are frequently offered for sale, the smaller male typically riding on the larger female's back). Boxer shrimps will sometimes eat other species of shrimp, especially in small tanks, so are best kept as the only shrimps in the aquarium unless the system is very large. They are robust enough to live with fishes that either prey on shrimps (such as small hawkfishes) or attack them for other reasons (some dottybacks, for example).

Several other *Stenopus* species are also imported occasionally, some larger and some smaller than the boxer shrimp. They require similar care to boxer shrimps. Only keep one *Stenopus* species (or a pair of one species) per tank, as inter-species territorial disputes can occur.

Pistol shrimps (*Alpheus* species)

Best known for sharing burrows in symbiotic relationships with a variety of gobies, pistol shrimps are unusual crustaceans that have a novel (and remarkable) way of capturing their prey (usually small crustaceans). Using a specially modified, enlarged claw, they produce a high-pressure pulse of water that is "shot" at their prey, stunning or even killing it. This is accompanied by a loud popping noise, which gives these shrimps their common name.

Some pistol shrimps are green or brown, but others are more attractively colored in shades of red, orange, yellow, pink, and purple. They are often rather

THE PISTOL SHRIMP STRIKE

Pistol shrimps generate the high-pressure pulse of water used to stun prey by snapping shut a specially modified claw. This results in water shooting out of the claw at a speed of over 62 miles per hour, which is impressive enough in itself, but not the whole story. The loud popping noise that accompanies the strike (which is easily heard outside the aquarium) is created by the collapse of a bubble, formed as the water jet is emitted, within the claw. Using high-speed cameras, scientists have discovered that as the bubble collapses, a tiny flash of light is emitted, in a process known as sonoluminescence (the conversion of sound energy into light). This is an extremely rapid process. The flash lasts no more than 10 billionths of a second, and is invisible to the human eye. Even more amazing, the wavelength of light emitted suggests that within the tiny, collapsing bubble, the temperature reaches phenomenal levels, approaching 9,000°F (5,000°C)! The small size and brief duration of the bubble means that this is of no practical significance—indeed the whole phenomenon just appears to be a side-effect of creating the rapid pulse of water—but it must be the highest temperature recorded within a living creature.

Left: *Pistol shrimps are excellent aquarium inhabitants, especially when kept with shrimp gobies, and some are very attractive.*

secretive when housed without a goby partner, but in tanks without large fishes that could be perceived as a threat, they can sometimes be quite extroverted. However, it is best to buy the shrimps with their partner gobies. Pairing up gobies and shrimps purchased separately is not always successful.

In the aquarium, pistol shrimps should be provided with a substrate bed of mixed sand and gravel pieces, at least in one area of the tank, that they can use to build a burrow. Constructing, maintaining, and remodeling elaborate

Below: The emerald crab (Mithrax sculptus) *is a useful algae grazer and one of the very few crabs that is safe in the reef aquarium.*

burrows is pretty much a full-time occupation for most pistol shrimps. They will feed on frozen crustaceans and sometimes dry foods, as well as hunting small creatures in the substrate. Pistol shrimps can be territorial, so unless you find a mated pair, or have a large tank, only keep one per aquarium.

Emerald crab (*Mithrax sculptus*)

The emerald crab *(Mithrax sculptus)* is a small spider crab, even though it looks very much like a true crab. As its name suggests it is bright green. It grows to about 2 inches across, and in the wild is found in the Caribbean. Unusually for a crab, it is mainly herbivorous, using its specially adapted claws to collect

algae. It is useful in the aquarium for the control of hair algae and bubble algae (*Valonia* species), which few other species apart from large tangs will tackle. It also eats fleshy macroalgae, and will scavenge other food remains. Emerald crabs are not particularly territorial among themselves, and are usually harmless to other invertebrates, and to fish. Except for a supply of algae, they do not seem to need any special conditions in the aquarium.

Hermit crabs

Small hermit crabs are not just interesting and attractive, but can be very beneficial in the reef aquarium. They can help to prevent algae

the available supply. They may also kill snails, extracting them from their shells. The snail itself may be eaten or discarded, but the hermit crab will use the shell.

Small hermits vary in their habitat preferences: some stay mainly on the sand, whereas others live on rocks or like to rest among the branches of corals.

Blue-legged hermit (*Clibanarius tricolor*)

The blue-legged hermit comes from sandy areas around reefs in the Caribbean, and is one of the smaller hermit crabs, growing to less than an inch. It has sky-blue legs, with the joints highlighted in red and yellow. It grazes algae and is particularly useful, as its small size means it can reach into smaller crevices than most other hermits.

Scarlet reef hermit (*Paguristes cadenati*)

This Caribbean species is a very popular hermit for the reef aquarium. It has bright scarlet legs and claws, and bright yellow eyes and grows to about 1.2

problems, both by grazing and by breaking up pockets of detritus as they scavenge for food. Very small hermits can get to places that other grazers, such as snails, cannot reach. Hermits also help to clear up uneaten fish food and dispose of any dead creatures.

When stocking hermits in a reef tank it is important not to add too many; they are generally inexpensive and it is tempting to add large numbers in an attempt to maximize their benefits. If the population of hermit crabs is too large for the available supply of algae or waste food, they may start to eat other things, such as small polyps and snails (whose shells they will then use). A sensible stocking level might be in the range of one hermit per 4 to 13 gallons of tank capacity, depending on the ultimate size of the crabs.

A key requirement of hermit crabs in the aquarium is a plentiful supply of empty shells of an appropriate size and type to use as they grow. It is best to provide a good variety of shells, as different hermit crabs prefer different types. Without enough spare shells, the hermits are more likely to fight over

Above: *Shrimps, here a fire shrimp, and small hermit crabs quickly scavenge any dead animals, such as this clownfish, before they decay and pollute the aquarium.*

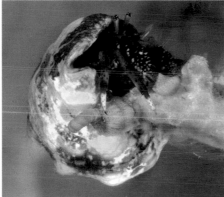

Above: *The tiny blue-legged hermit is a colorful addition to the reef tank and a useful algae grazer.*

Right: *Like all small hermit crabs, scarlet reef hermits are fascinating to observe, as well as attractive and useful.*

89

inches. In the wild it lives in sandy areas close to reefs. In the aquarium it grazes algae, as well as eating scraps of fish food. Occasionally, this species becomes yellow or orange rather than scarlet after molting, but generally reverts to its original color at the next molt.

Red-legged hermit
(*Paguristes digueti*)

This is a similar species to the scarlet reef hermit, but not so vividly colored (its legs are dull red rather than the bright scarlet of *P. cadenati*). The red-legged hermit comes from the Gulf of California. It grows larger than the scarlet reef hermit, to about 2.3 inches across the legs, and is more of a general scavenger than an algae grazer.

Blue-knuckle hermit
(*Calcinus elegans*)

This Pacific hermit is spectacularly colored, with vivid electric blue stripes on the black legs, bright blue eye

stalks, and orange antennae. There is a Hawaiian color variant, with orange rather than blue bands on the legs (sometimes sold as a Halloween hermit). *C. elegans* grows to about 2 inches across, and lives in intertidal regions of rocky shores. In the aquarium it is an enthusiastic algae grazer, preferring to feed from rocks. It also eats a range of other foods.

Blue-eyed hermit
(*Calcinus laevimanus*)

Calcinus laevimanus has large black claws tipped in bright white, orange legs and eye-stalks, and bright blue eyes. It grows to about 1.5 inches across and comes from Florida and the Caribbean. In the reef aquarium, it is an excellent algae grazer.

Orange-striped hermit
(*Trizopagurus strigatus*)

This species is both larger (it grows to about 2 inches across) and usually

much more expensive than most other small hermit crabs. Although not a useful algae grazer in the aquarium, it is a good general scavenger. The legs and claws of this very attractive animal have orange bands over a bright yellow background. The rather flattened body suits this hermit's preferred shells, those of venomous cone shells. It is found in the Red Sea and Indo-Pacific Ocean.

Left: *The blue-knuckle hermit* (Calcinus elegans) *is larger than most other small hermits. A form with orange markings on the legs is occasionally available.*

Left: *The orange-striped hermit* (Trizopagurus strigatus) *is notable for living in the empty shell of the cone shell. It is a good general scavenger.*

Above: Calcinus laevimanus *has a left claw much larger than the right one. It is an effective algae grazer in the reef aquarium.*

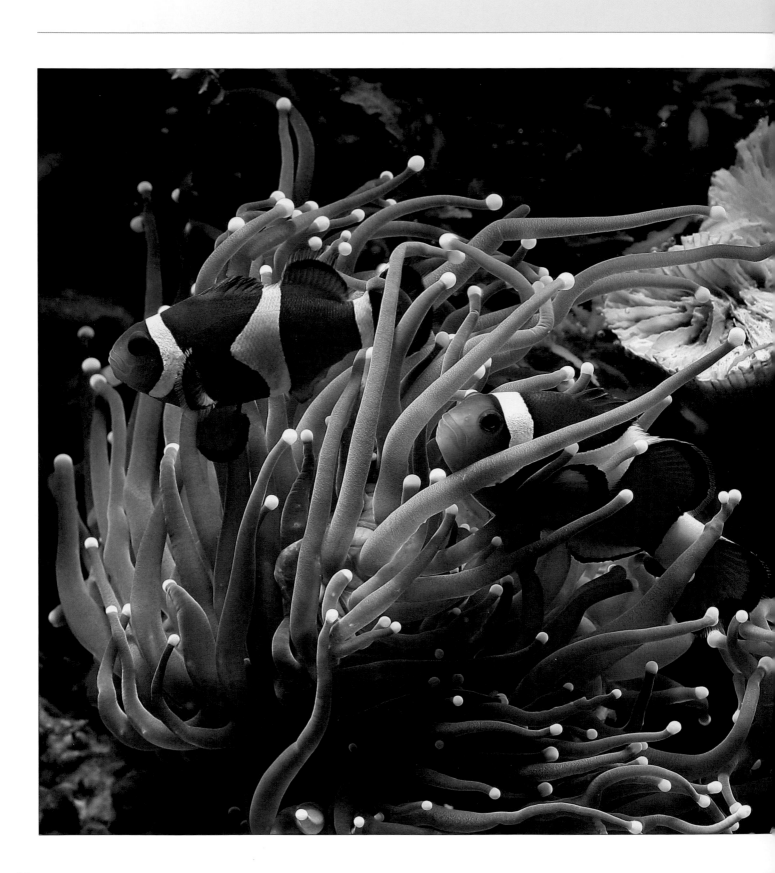

Chapter Four

Fishes for the reef aquarium

It is entirely possible to keep a reef aquarium without fishes, but for most enthusiasts they are key members of the reef tank community. They bring an extra dimension to the aquarium with their movement and often dazzling colors, and in some cases become real pets, developing a relationship with their keepers. On a more mundane note, some fishes can help in controlling various pests and problems.

A reef aquarium makes some special demands of its fishes. In an aquarium devoted to the culture of corals and other invertebrates, fish that eat such creatures are obviously undesirable. This initially rules out many popular aquarium species, including, among others, most butterfly fishes, angelfishes, puffer fishes, and triggerfishes.

Some of these are a threat to a wide range of invertebrates, while others have more specific appetites. Indeed, many fishes in the latter category can be accommodated in a reef aquarium as long as the specific invertebrates likely to be eaten are excluded from the tank. This approach is enabling aquarists to keep a much wider range of fishes with invertebrates, but our knowledge of the precise combinations of creatures that will cohabit in the aquarium is still very much in its infancy.

Common clownfishes frolic among the tentacles of a Euphyllia glabrescens *stony coral.*

Tangs or surgeonfishes

The tangs or surgeonfishes (the two terms can be used interchangeably) deserve a special focus, as they are among the few large fishes that can be kept with invertebrates in almost complete safety; most species are herbivores, grazing algae, and some also eat detritus. They are often very colorful, lively fishes, and many are such efficient grazers that they can sometimes eliminate algae problems. There are many species of tangs, but those featured here are among the easiest to keep. For these species, their aquarium husbandry is not particularly difficult, given a basic understanding of their needs.

Choosing tangs

When buying tangs, it is best to select relatively small, young individuals, as these adapt better to aquarium life than large adults. Avoid very small tangs; for most species the best size to buy is in the range of 2–4 inches.

Compatibility

Tangs generally get along fine with fish that they do not regard as competitors for the supply of algae. Planktivorous species, substrate-feeding gobies, etc., are usually tolerated or ignored.

SELECTING FISHES FOR THE REEF AQUARIUM

In making a selection of reef fishes for this book, we have taken a conservative approach. The species described here are usually safe with just about all invertebrates, although it is worth noting that individual fishes vary to some degree in their behavior, so it is possible (if unfortunate) to find rogue individuals of most species.

A less obvious characteristic that is desirable in reef aquarium fishes is hardiness and resistance to disease. Although reef aquariums provide a very high-quality environment for fishes (because invertebrates have stricter environmental requirements than fishes), an inherently low susceptibility to disease is a great advantage. Most treatments for fish diseases are toxic to invertebrates and so cannot be used in the reef aquarium. Similarly, it is important that fishes adapt easily to aquarium life and are easy to feed. Transferring a fish that has a parasitic infection or is having problems feeding to a treatment tank is even more difficult in the reef tank than in a system devoted exclusively to fishes, because catching fish in an aquarium full of corals is no easy matter. With this in mind, the fish selection presented here family by family focuses on some of the most attractive, reliable, and easily available species. They usually do very well in the aquarium and should prove to be hardy and disease resistant, provided you observe a few simple precautions.

Mixing tangs with other tangs is trickier, but can work. Large tanks, an abundant supply of food, and plenty of hiding places can help, as can keeping

Below: In large tanks, yellow tangs can be kept in groups and mixed with other Zebrasoma *species, such as the purple tang in this aquarium—although these species would never meet in the wild.*

species of different appearance (ideally in size, shape and color), and adding the less territorial fish first. If you add a tang to a tank where one is already resident, you can expect some initial sparring, but usually this settles down quickly. Of the species described here, only the yellow tang (*Zebrasoma flavescens)* can be reliably kept with others of its own species. Provide a large tank (at least 72 × 18 × 22 inches) and introduce groups of at least five individuals simultaneously.

Feeding tangs

Diet is probably the most critical factor for success with tangs. In the wild, tangs graze algae most of the day, eating plenty but extracting relatively little nutrition from each mouthful, so they need large quantities of food. Tangs also appear to require abundant supplies of vitamins, particularly vitamin C.

live algae they will eat. Even if you do have the right algae, the tang will probably deplete it very quickly.

Health problems

As well as HLLE, tangs as a family are susceptible to *Cryptocaryon irritans* (white spot) infections. The species described here are not among the most susceptible tangs, and avoiding stress, for example caused by poor water conditions, bullying by tankmates, or temperature swings, can minimize the chances of infection. Quarantining tangs, either under low-salinity conditions or in normal sea water, is highly recommended. Quarantine procedures are described on page 163. If your fish do become infected, treatment with either low-salinity water or copper (in both cases in a separate hospital tank) is usually effective.

Without proper feeding, tangs become thin, their colors fade, they are more vulnerable to infection, and may develop head and lateral line erosion (HLLE), which appears to be related to vitamin C deficiency (and can be treated by feeding a proper diet).

The best staple foods in the aquarium are commercial algae-based products. Offer sheets of toasted algae, either nori (as used for sushi) or similar products packaged for the aquarium trade. Other forms of dried seaweed are available and worth trying if you can find them. Algae-based flake and pellet foods are also beneficial, as is frozen brine shrimp enriched with *Spirulina*. The wide range of algae-based foods available means that there is no need to feed tangs with

Above: *Powder blue tangs* (Acanthurus leucosternon), *here in the Indian Ocean, are difficult to keep in the reef aquarium. They are very susceptible to white spot, which can be hard to manage in this setting.*

lettuce or other vegetables, which are nutritionally inferior.

Ideally, tangs need to be fed several times a day. A good method is to provide dried algae sheets using a food clip, and to replenish the supply a couple of times a day, as well as using other dried and frozen foods.

Tangs will eat algae growing in the aquarium, but it is not a good idea to rely on this as a sole food source. Although they readily accept prepared foods, tangs can be fussy about which

DEFENSIVE WEAPONS

Tangs or surgeonfishes are so called because of their defensive weapons—specially modified scales, that lie on each side of the caudal peduncle, where the tail fin joins the body. These are very sharp, scalpel-like structures that can inflict serious wounds on other fishes. In most species they can be retracted when not in use. In some species they are venomous. Generally speaking, aquarium keepers are not at risk when working in their tanks, but great care is needed when netting tangs and they should not be handled if at all possible.

Zebrasoma tangs

Zebrasoma species have short, high bodies, with tall dorsal fins. They typically have long noses and feed on filamentous algae and seaweeds. In the aquarium, *Zebrasoma* species are among the best tangs for controlling nuisance algae; all species eat most forms of filamentous algae and larger individuals will often eat bubble algae. *Zebrasoma* tangs are also probably the least susceptible of the whole family to white spot; the large sailfins, *Z. veliferum* and *Z. desjardinii*, are particularly good in this respect.

In addition to the species described here, you may also come across two other beautiful but high-priced species: *Z. rostratum*, the black tang, and the very rare and astonishingly expensive *Z. gemmatum*.

YELLOW TANG
Zebrasoma flavescens

This fish is a beautiful daffodil-yellow color, almost luminous in the best specimens, with the scalpels outlined in bright white. It is the smallest *Zebrasoma* tang and generally peaceful with other species.

Maximum size: 8 inches

Minimum tank size: 48 × 18 × 20 inches (75 gallons)

Geographic origins: Pacific Ocean from Hawaii to the Ryukyu Islands.

The "scalpels," specially modified scales that give surgeonfishes their name, are very obvious in yellow tangs.

SCOPAS TANG
Zebrasoma scopas

Although more soberly colored than the other *Zebrasoma* species, *Z. scopas* is still an attractive fish and an excellent algae grazer. It is quite an aggressive species, and needs to be one of the last fishes added to the aquarium.

Maximum size: 8 inches

Minimum tank size: 48 × 18 × 20 inches (75 gallons)

Geographic origins: Wide range, from East Africa, Indian Ocean, and Pacific Ocean.

PURPLE TANG
Zebrasoma xanthurum

A stunning, deep purple fish, with yellow tail and pectoral fins, nostrils and lips. It is one of the more aggressive *Zebrasoma* tangs and as such should be one of the last fishes added to a mixed community. It is usually shipped from the Red Sea and so tends to be expensive, but is well worth its high price, as it is usually very hardy and a reliable aquarium species.

Maximum size: 8.5 inches

Minimum tank size: 48 × 24 × 24 inches (120 gallons)

Geographic origins: Red Sea, Persian Gulf, and Western Indian Ocean.

PACIFIC SAILFIN TANG
Zebrasoma veliferum

This species and the Indian Ocean/Red Sea sailfin tang (*Zebrasoma desjardinii*) are very similar, especially when young; they are sometimes considered to be geographical subtypes of the same species. They have particularly large fins (most prominent in small individuals), hence their common names. They are the biggest of the *Zebrasoma* tangs and require large tanks. They generally have a peaceful disposition, except where members of their own species are concerned, and are among the hardiest of the tang family. They are very relaxed in the aquarium, unlike some other tangs that can be rather jittery. Their only negative point is their large size. They can change color dramatically according to their mood.

Maximum size: 16 inches

Minimum tank size: 72 × 24 × 24 inches (180 gallons)

Geographic origins: *Z. veliferum*: Pacific Ocean, from Hawaii to Indonesia, southern Japan to New Caledonia. *Z. desjardinii*: Indian Ocean and Red Sea.

Right top: *Adult Z. desjardinii.*
Right: *Juvenile Z. desjardinii.*

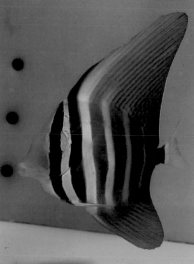

Above: *Subadult Z. veliferum.*

Ctenochaetus **tangs**

The *Ctenochaetus* species are sometimes referred to as bristletooth tangs. They have a different method of feeding from *Zebrasoma* and *Acanthurus* species, which have jaws that are adapted to snipping off pieces of filamentous algae or macroalgae. *Ctenochaetus* tangs, in contrast, have lips lined with brushlike teeth that can scrape microalgae films from the substrate. Many *Ctenochaetus* species also include a high proportion of detritus in their diet. In tanks that are sufficiently large, the combination of *Zebrasoma* and *Ctenochaetus* tangs makes an ideal team for algae control, thanks to their complementary grazing activities.

    Ctenochaetus species are generally peaceful (except towards their own species) and are sometimes picked on by more aggressive tangs. They are rather shy by tang standards, especially when first added to the aquarium, and are somewhat more susceptible to infections than *Zebrasoma* species. In addition to those described here, others of the nine *Ctenochaetus* species are also imported occasionally.

The silver spot tang (C. truncatus) requires very similar care to the kole tang.

KOLE, OR YELLOW-EYE, TANG
Ctenochaetus strigosus

The kole tang and the spotted bristletooth, or silver spot, tang *(C. truncatus)* are among the smallest tangs and both species generally adapt well to aquarium life. The kole tang is dark brown with lighter "pinstripes" and bright gold eyes. The spotted bristletooth is very similar, but its flanks are covered with lighter spots instead of pinstripes. Some taxonomists have regarded *C. truncatus* as a variant of *C. strigosus*, but it is currently considered to be a separate species. Both species can lighten or darken their colors quite markedly according to mood, but the color can fade on prolonged exposure to copper, low levels of which are sometimes used in dealers' tanks. They are usually very peaceful species, ignoring tankmates.

Maximum size: 6 inches

Minimum tank size: 48 × 18 × 20 inches (75 gallons)

Geographic origins: Kole tang: Central Pacific Ocean, especially Hawaii and Johnston Island. Spotted bristletooth: Indian Ocean.

CHEVRON TANG
Ctenochaetus hawaiiensis

This is one of the larger *Ctenochaetus* species and usually sold in its stunningly beautiful juvenile form—bright orange with purplish-blue markings in a vague chevron pattern, often with vivid purple-blue fins. However, its color changes with age, becoming almost black with lighter pinstripes. This process takes several years and it remains a handsome fish, even in adult form. It is a deep water species (and so is usually quite expensive), but adapts very well to brightly lit reef tanks. It is a very peaceful species.

Maximum size: 10 inches

Minimum tank size:
72 × 18 × 22 inches
(125 gallons)

Geographic origins:
Pacific Ocean, from Micronesia to Hawaii and Pitcairn Island.

Very colorful juvenile chevron tangs grow into more soberly colored but still handsome adults (below).

ORANGE-TIPPED BRISTLETOOTH
Ctenochaetus tominiensis

The peaceful orange-tipped bristletooth requires very similar care to other *Ctenochaetus* tangs. It is a subtly attractive species, with a velvety brown body, bright orange edges, and tips to the dorsal and anal fins and a white tail fin.

Maximum size: 6 inches

Minimum tank size: 48 × 18 × 20 inches
(75 gallons)

Geographic origins: Western and Central Pacific Ocean, from Indonesia to Fiji

Acanthurus surgeonfishes

Acanthurus is the largest genus of tangs and surgeonfishes, with 38 species distributed throughout the world's warm seas. These include some aquarium favorites, but many *Acanthurus* species are challenging to keep due to size, disease susceptibility, extreme territoriality, strict habitat requirements, or any combination of these factors (all of them in some cases). The species described here are a selection of the more reliable, easier-to-keep *Acanthurus* tangs.

BLUE TANG
Acanthurus coeruleus

The only Atlantic fish among our selection of tangs is an attractive, robust fish and one of the better *Acanthurus* species for algae control. It is quite territorial, especially in smaller tanks. When young, it is bright yellow with a blue ring around the eye. It passes through a phase where it is pale beige, but then as an adult develops a lovely rich blue color (initially with a yellow tail, but this turns blue eventually), which it can vary according to mood.

Maximum size: 9 inches

Minimum tank size: 72 × 18 × 22 inches (125 gallons)

Geographic origins: Western Atlantic Ocean, particularly the Caribbean, Bahamas, and Florida.

ORANGESHOULDER SURGEONFISH
Acanthurus olivaceus

This large species requires a big aquarium. *A. olivaceus* is another fish that undergoes dramatic color changes as it grows. Juveniles are solid bright yellow, gradually developing the characteristic orange patch and two-tone body with age. This species feeds on algae and detritus from sand beds. It is usually one of the more peaceful *Acanthurus* species, but some individuals can be territorial.

Maximum size: 14 inches

Minimum tank size: 72 × 24 × 24 inches (180 gallons)

Geographic origins: Western Pacific Ocean, over a wide range.

Left: *The juvenile* A. olivaceus *is very different from the adult.*

MIMIC, OR CHOCOLATE, SURGEON
Acanthurus pyroferus

Both *Acanthurus pyroferus* and the Indian Ocean mimic surgeon (*A. tristis*) are excellent aquarium species and two of the most peaceful of the *Acanthurus* species. They are notable for the fact that when young they mimic a number of *Centropyge* dwarf angelfishes: *C. flavissimus* (the lemonpeel angel) and *C. vroliki* (half-black, or pearlscale, angel) for *A. pyroferus*, and *C. eibli* for *A. tristis*. The color scheme and fin shapes of the juvenile tangs match those of the angels. Adults have very different color schemes to juveniles. *A. pyroferus* varies between yellow shading to brown on the rear half of the body and rich chocolate brown, in both cases with striking facial markings, while *A. tristis* is a lighter brown with darker fins and similar facial markings to *A. pyroferus*. The tall fin in adults of both species is lyre shaped, in contrast to the convex tails of the juveniles. The mimicry of dwarf angels is probably a protective mechanism. In juvenile tangs the scalpels are relatively small and probably of limited use in defense against predators. *Centropyge* angels, however, are armed with formidable spines on the gill covers and make a spiky mouthful. With age and increasing size, the tangs' own defenses become more effective, and the mimicry becomes unconvincing as the tangs grow much larger than the angelfishes.

Maximum size: 10 inches

Minimum tank size: 72 × 18 × 22 inches (125 gallons)

Geographic origins: *Acanthurus pyroferus:* Indo-Pacific Ocean; *Acanthurus tristis:* Indian Ocean.

Right: *Juvenile* A. pyroferus *mimicking* Centropyge flavissimus *(top), similar to juvenile* A. olivaceus *(bottom).*

CONVICT SURGEON
Acanthurus triostegus

This is another laid-back *Acanthurus* species, and an excellent algae grazer for the reef aquarium. In the wild it sometimes forms large shoals, but in the aquarium it is best kept as a single individual, although it mixes well with other tangs. Although less colorful than some other members of the family, a well-fed, healthy specimen is a very attractive fish, with dark bars over a shimmering metallic base color.

Maximum size: 10 inches

Minimum tank size: 72 × 18 × 22 inches (125 gallons)

Geographic origins: Indo-Pacific Ocean to Eastern Pacific, including Galapagos Islands and Gulf of California.

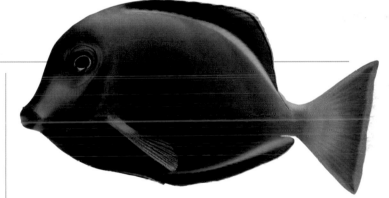

POWDER BROWN, OR GOLDRIM, TANG
Acanthurus japonicus

The powder brown tang is a beautiful fish, with a brown body (the fish can vary its color intensity) accented by brightly colored fins. This species is a bit more of a typical *Acanthurus* than the others described here. It can be quite territorial and is more prone to white spot, especially when settling into a new aquarium, than the other species described here. However, this lovely fish is much easier to keep than many of its relatives.

Maximum size: 8 inches

Minimum tank size: 48 × 24 × 24 inches (120 gallons)

Geographic origins: Indo-West Pacific.

Angelfishes

Angelfishes are among the all-time favorite fishes for one very good reason—many are very beautiful, even if some are hard to keep. For the reef aquarium keeper, choosing angelfishes is a tricky business, as many angels may eat sessile invertebrates. The species featured here are usually reef safe.

Centropyge angels

Centropyge angels typically grow to about 3 to 4 inches. They vary tremendously in their ease of aquarium keeping and their safety with sessile invertebrates. The species described are all reliable in the aquarium and usually leave corals alone. However, they will eat sponges and tubeworms on living rock, as well as algae and detritus. Occasional rogue specimens sometimes bother fleshy, large-polyp stony corals or tridacnid clams.

Once established in the aquarium, these species will eat most foods they are offered, but it is a good idea to provide algae-based foods. They also appreciate having diatom films and detritus to graze on. They like plenty of cover in the aquarium, which will encourage them to be bolder. They will not usually tolerate others of their own species, except in very large tanks. Most *Centropyge* angels do not bother other fishes, but the cherub angel *(C. argi)* can be quite territorial, especially in small aquariums.

CORAL BEAUTY
Centropyge bispinosus

Maximum size: 4 inches

Minimum tank size: 36 × 18 × 18 inches (50 gallons)

Geographic origins: Indo-Pacific Ocean, from East Africa to the Tuamoto Islands.

FLAMEBACK ANGEL
Centropyge acanthops

Maximum size: 3 inches

Minimum tank size: 36 × 18 × 16 inches (40 gallons)

Geographic origins: Western Indian Ocean.

CHERUB ANGEL
Centropyge argi

Maximum size: 3 inches

Minimum tank size: 36 × 18 × 16 inches (40 gallons)

Geographic origins: Western Atlantic Ocean, Caribbean, and Gulf of Mexico.

Genicanthus angels

Genicanthus angels are almost exclusively planktivorous and are the safest angel family for the reef aquarium. They are unusual in that males and females are usually different in color and pattern—sometimes very different. They can be kept in pairs, or one male with multiple females, provided that all individuals are introduced together. All *Genicanthus* angels start life as females, with dominant individuals becoming males as they grow.

Once established in the aquarium they are usually hardy and not particularly territorial, although they occasionally chase other planktivores. They will eat most foods, but frozen crustaceans such as brine shrimp, mysis, and krill are ideal. Many species are caught in deep water, and when buying them, you should watch for decompression injuries, which usually manifest themselves as an inability to hold position in the water column. *Genicanthus* angels are active swimmers and need plenty of open water.

JAPANESE SWALLOWTAIL ANGEL
Genicanthus semifasciatus

Maximum size: 8 inches
Minimum tank size: 48 × 24 × 24 inches (120 gallons)
Geographic origins: Western Pacific, Southern Japan to Taiwan and the Philippines.

LAMARCK'S ANGEL
Genicanthus lamarck

Maximum size: 10 inches
Minimum tank size:
72 × 18 × 22 inches (125 gallons)
Geographic origins: Indo-West Pacific, Malaya to Vanuatu.

♀

BLACKSPOT ANGEL
Genicanthus melanospilos

Maximum size: 7 inches
Minimum tank size:
48 × 24 × 24 inches (120 gallons)
Geographic origins: Western Pacific, Indonesia to Fiji.

♀

♀

BELLUS ANGEL
Genicanthus bellus

Maximum size: 7 inches
Minimum tank size:
48 × 24 × 24 inches (120 gallons)
Geographic origins: Eastern Indian Ocean, Pacific Ocean.

♂ Male ♀ Female

♂

Right: *As in most* Genicanthus *species, male and female blackspot angels look very different.*

Gobies

The goby family is a very large one that includes many species that make excellent inhabitants for the reef aquarium. The species described here are just a taster of the many wonderful gobies that you can keep.

Shrimp gobies

Many gobies live in symbiotic relationships with pistol shrimps. The shrimps dig burrows that they share with the fishes (often a pair of gobies with one or two shrimps), while the fishes act as lookouts for predators and catch food, the scraps of which are eaten by the shrimps. The gobies do not need their shrimp partners in the aquarium, although if you can buy both together it is fascinating to observe the relationship. With or without shrimps, the gobies do need a bed of sand in the aquarium (preferably with a mixture of different particle sizes) to dig burrows. Many of these gobies are territorial towards their own species and other shrimp gobies. Most are harmless to invertebrates, although species with larger mouths (the yellow watchman goby, for example) may eat very small shrimps.

BARBER POLE GOBY
Stonogobiops nematodes

Maximum size: 2.4 inches
Minimum tank size: 24 × 12 × 16 inches (20 gallons)
Geographic origins: Western Pacific Ocean, Philippines to Samoa.

CLOWN GOBY
Stonogobiops yasha

The prominent dorsal fin ray is used as a signal device.

Maximum size: 2 inches
Minimum tank size: 24 × 12 × 16 inches (20 gallons)
Geographic origins: Western Pacific Ocean.

TANGERINE-STRIPED GOBY
Amblyeleotris randalli

Maximum size: 5 inches
Minimum tank size: 48 × 18 × 20 inches (75 gallons)
Geographic origins: Western Pacific Ocean.

YELLOW WATCHMAN GOBY
Cryptocentrus cinctus

Maximum size: 3 inches
Minimum tank size: 36 × 12 × 16 inches (30 gallons)
Geographic origins: Western Pacific Ocean.

Neon gobies

The neon gobies and their relatives are diminutive fishes, ideal for smaller reef tanks. Some species, such as the neon and gold neon gobies, act as cleaners, picking parasites off larger fishes. They are not territorial toward other fishes, but may fight with their own species in smaller tanks. Some species will spawn in the aquarium. In view of their small size, it is best not to house them with predators—even small ones.

NEON GOBY
Elacatinus oceanops

Maximum size: 2 inches
Minimum tank size: 24 × 12 × 16 inches (20 gallons)
Geographic origins: Western Atlantic Ocean, Florida to Belize.

Other gobies

Here are a few more varieties of goby to consider.

GOLD NEON GOBY
Elacatinus randalli

This species is very similar to the neon goby (*E. oceanops*), but instead of a neon blue stripe along the flanks, it has a metallic gold streak. It is another ideal species for the small reef tank.

Maximum size: 1.8 inches
Minimum tank size: 24 × 12 × 16 inches (20 gallons)

Geographic origins: Western Central Atlantic Ocean; Puerto Rico to Venezuela.

RED-HEADED GOBY
Elacatinus puncticulatus

Another neon goby relative, this species has bright neon red-and-blue stripes on the head, and a series of dark patches running along the flanks of its almost transparent body.

Maximum size: 1.8 inches
Minimum tank size: 24 × 12 × 16 inches (20 gallons)

Geographic origins: Eastern Central Pacific Ocean.

STRIPED SHRIMP GOBY
Amblyeleotris fasciata

This species is typical of *Amblyeleotris* gobies. It lacks the spectacular dorsal fin of *A. randalli*, but has a series of broad crimson bands along the body and blue highlights in most of the fins.

Maximum size: 3 inches
Minimum tank size: 35 × 12 × 16 inches (30 gallons)

Geographic origins: Western Pacific Ocean.

Coral gobies

Gobiodon gobies live among the branches of stony corals, such as *Acropora* species. Being very small, they must be kept with species that will not eat them. They must also be fed suitably sized foods. They are territorial toward their own species, but in large tanks with sufficient coral colonies it is possible to keep groups. Pairs may form and spawn among the coral branches.

YELLOW CORAL GOBY
Gobiodon okinawae

Maximum size: 1.4 inches
Minimum tank size:
24 × 12 × 16 inches (20 gallons)
Geographic origins: Western Pacific Ocean.

CITRON CORAL GOBY
Gobiodon citrinus

Maximum size: 2.4 inches
Minimum tank size:
24 × 12 × 16 inches (20 gallons)
Geographic origins: Red Sea, Indian and Western Pacific Oceans.

GREEN CORAL GOBY
Gobiodon histrio

Maximum size: 1.4 inches
Minimum tank size:
24 × 12 × 16 inches (20 gallons)
Geographic origins: Red Sea, Indian and Western Pacific Oceans.

A small goby with very attractive markings.

Firefishes and dartfishes

The dartfish (family Microdesmidae) is closely related to the gobies and includes some excellent species for the reef aquarium. These are planktivorous fishes that hover close to the substrate. They are generally disease resistant and will eat a range of dried and frozen foods. They appreciate plenty of cover into which they can retreat when frightened; they can be quite nervous, so also prefer quiet companions. Keep the tank well covered, as most species will jump when frightened. House the firefishes singly or in pairs. Dartfishes can be kept in groups (including mixed-species groups), although pairs will often form.

PURPLE FIREFISH
Nemateleotris decora

Maximum size: 3.5 inches
Minimum tank size:
36 × 18 × 16 inches (40 gallons)
Geographic origins: Indian and Pacific Oceans.

FIREFISH
Nematelcotris magnifica

Maximum size: 3.5 inches
Minimum tank size:
36 × 18 × 16 inches (40 gallons)
Geographic origins: Indian and Pacific Oceans.

ZEBRA DARTFISH
Ptereleotris zebra

Maximum size: 4.7 inches
Minimum tank size:
48 × 18 × 20 inches (75 gallons)
Geographic origins: Indian and Pacific Oceans.

SCISSORTAIL DARTFISH
Ptereleotris evides

Maximum size: 5.5 inches
Minimum tank size:
48 × 18 × 20 inches (75 gallons)
Geographic origins: Red Sea, Indian and Pacific Oceans.

Blennies

The blennies are a large, diverse family of fishes, many of which have unusual and fascinating lifestyles. They range from algae grazers, through planktivores and predators of corals, to species that are essentially parasites, nipping skin and scales from other fishes.

Blennies that prey on other fishes in this way are obviously undesirable reef aquarium inhabitants, as are those that graze on corals. Other blennies are very territorial and some of the algae grazers may turn to grazing on corals (particularly fleshy, large-polyp stony corals) if there is insufficient algae in the tank. Blennies also vary considerably in their hardiness in the aquarium, but those described here are generally robust species that do well and are harmless to invertebrates.

ALGAE BLENNY
Salarias fasciatus

The algae blenny looks much like the kind of blenny that is familiar from seaside rockpools. It is intricately patterned but not brightly colored, although its flamboyant fins give it a spectacular appearance. As its name suggests, it feeds by grazing on filamentous algae and diatom films, and needs a supply of algae-based foods. It can be quite territorial with its own species and similar-looking fishes.

Maximum size: 5.5 inches
Minimum tank size: 48 × 18 × 20 inches (75 gallons)
Geographic origins: Red Sea, Indian and Pacific Oceans.

MIDAS BLENNY
Ecsenius midas

This is a predominantly planktivorous member of a genus mainly made up of algae grazers. It usually does not bother corals. Its color is variable, depending on where it is collected, although it can also change its pattern according to mood. This species adapts very well to the aquarium and often becomes a real pet. It accepts most foods; a diet of frozen crustaceans and some algae-based flake suits it well. It is not particularly territorial, except when kept in small aquariums.

Maximum size: 5 inches
Minimum tank size: 48 × 18 × 20 inches (75 gallons)

Geographic origins: Red Sea, Indian and Pacific Oceans.

Meiacanthus blennies

Meiacanthus species are unlike most blennies in that they have very well-developed swimbladders and can therefore hover easily in the water column. Their elegant swimming contrasts sharply with the rather clumsy efforts of other blennies. Swimming out in the open would leave them vulnerable to predation, but they have evolved a defense mechanism in the form of a venomous bite. Once bitten, most predators quickly let go of these blennies and learn to avoid them in future. To humans, the bites are usually no worse than bee stings, but it is still worth taking care to avoid them in case you happen to be allergic to the venom. Fortunately, these are not aggressive fishes.

In the wild, these blennies catch zooplankton and hunt small crustaceans on the substrate. Some *Meiacanthus* species can be difficult to feed in the aquarium, but the species shown here usually adapt easily to aquarium foods, eating both frozen crustaceans and dried foods with relish. *Meiacanthus* species are harmless to sessile invertebrates.

YELLOWTAIL FANG BLENNY
Meiacanthus atrodorsalis

Maximum size: 4 inches
Minimum tank size: 48 × 18 × 20 inches (75 gallons)
Geographic origins: Western Pacific Ocean.

STRIPED FANG BLENNY
Meiacanthus grammistes

Maximum size: 4 inches
Minimum tank size: 48 × 18 × 20 inches (75 gallons)
Geographic origins: Western Pacific Ocean.

Fang blennies are so-called on account of their large canine teeth.

MIMICRY IN BLENNIES

Mimicking other fishes is a blenny speciality. One species (the midas blenny) mimics Pseudanthias wreckfishes in order to enjoy the protection of swimming in their shoals; several blenny species mimic other blennies (harmless species imitate the venomous Meiacanthus blennies in order to deter attacks by predators); and some blennies even mimic cleaner wrasses, in order to be allowed to approach larger fishes without the risk of being eaten. They then betray the larger fishes' trust by biting off skin, scales, or pieces of their fins.

Basslets, grammas, forktails, and dottybacks

Most groupers and basses are too large to keep in a home aquarium, but they have many smaller relatives that make excellent fishes for the reef tank. These include some dazzlingly colorful species. Most of these are small, hardy, easy to keep, and will accept most foods, with small frozen crustaceans being favorites. Some species can be kept in pairs or small groups in larger tanks, and several spawn frequently in the aquarium. A few species are regularly available in tank-bred form.

Basses (*Serranus* species)

Serranus species are attractive miniature basses from the Tropical Atlantic and Caribbean. Both species featured here are harmless to sessile invertebrates but may eat small shrimps or very small fishes. The chalk bass can be kept in small groups (all must be added simultaneously), but the harlequin bass is more territorial and should be kept singly.

CHALK BASS
Serranus tortugarum

Maximum size: 3 inches
Minimum tank size:
36 × 18 × 16 inches (40 gallons)
Geographic origins: Southern Florida and Caribbean.

HARLEQUIN BASS
Serranus tigrinus

Maximum size: 4 inches
Minimum tank size: 48 × 18 × 20 inches (75 gallons)
Geographic origins: Western Atlantic Ocean and Caribbean from northern South America to Bermuda.

Stock single specimens of the hardy harlequin bass.

Grammas (*Gramma* species)

The grammas are dazzling little fishes, perfect for the reef aquarium. They appreciate having caves and holes to hide in, and if provided with an overhanging rock or a cave roof, will often orient themselves upside-down beneath it. They can be quite territorial about their chosen home. In the reef aquarium, grammas are only a threat to the smallest shrimps. The royal gramma can be kept in groups in larger tanks, but the blackcap gramma is best kept singly.

ROYAL GRAMMA
Gramma loreto

Maximum size: 3 inches
Minimum tank size: 36 × 18 × 16 inches (40 gallons)
Geographic origins: Bermuda, Bahamas, and Central America to northern South America.

BLACKCAP GRAMMA
Gramma melacara

Maximum size: 4 inches

Minimum tank size: 48 × 18 × 20 inches (75 gallons)

Geographic origins: West Indies, including the Bahamas, and Central America.

Forktails (*Assessor* species)

Forktails are small, peaceful fish that pose no threat to any invertebrates in the aquarium. They can be kept in small groups: the best approach to this is to add one larger individual and two or more smaller ones, to replicate the usual situation in the wild. They like to have plenty of cover in the aquarium, particularly large caves, where they will hover close to the roof (sometimes upside-down).

YELLOW FORKTAIL
Assessor flavissimus

Maximum size: 2 inches

Minimum tank size: 24 × 12 × 16 inches (20 gallons)

Geographic origins: Western Central Pacific Ocean: Great Barrier Reef.

BLUE FORKTAIL
Assessor macneilli

Maximum size: 2.3 inches

Minimum tank size: 24 × 12 × 16 inches (20 gallons)

Geographic origins: Western Pacific Ocean: Great Barrier Reef and New Caledonia.

111

Dottybacks (*Pseudochromis* species)

The dottybacks (family Pseudochromidae) are colorful fishes that include some near-perfect reef aquarium species, as well as others (particularly the larger *Labracinus* and *Ogilbyina* species), which are highly aggressive predators of smaller fishes and crustaceans and can become real terrors.

Here, the focus is on the small, often vividly colored members of the genus *Pseudochromis*. These vary in temperament; some are relatively peaceful and can be kept in groups or pairs (*P. fridmani*, *P. sankeyi*, and *P. springeri* fall into this category) and pose little threat to invertebrates (other than small bristleworms, which they relish eating). Others, such as *P. aldabraensis*, *P. diadema*, *P. paccagnellae* and *P. porphyreus*, are highly territorial and need to be kept with robust companions. The latter are not just a threat to other fishes: these species may attack and kill shrimps, such as cleaner shrimps (although better armored crustaceans, such as boxing shrimps, are usually safe).

Pseudochromis species prefer tanks with plenty of cover in the form of caves and crevices among rocks and corals. They vary somewhat in body shape: some are long, slender fishes with flowing fins (*P. aldabraensis*, *P. fridmani*, *P. sankeyi*, and *P. springeri*, for example), whereas others are much stockier with less flamboyant finnage (*P. diadema*, *P. paccagnellae*, and *P. porphyreus*). Some taxonomists have placed the latter type into a separate genus, *Pictichromis*.

ORCHID DOTTYBACK
Pseudochromis fridmani

A peaceful species that poses no threat to invertebrates.

Maximum size: 2.7 inches
Minimum tank size: 36 × 18 × 16 inches (40 gallons)
Geographic origins: Red Sea.

SUNRISE DOTTYBACK
Pseudochromis flavivertex

Maximum size: 2.7 inches
Minimum tank size:
36 × 18 × 16 inches (40 gallons)
Geographic origins: Red Sea, Western Indian Ocean.

SANKEY'S DOTTYBACK
Pseudochromis sankeyi

Maximum size: 2.7 inches
Minimum tank size:
36 × 18 × 16 inches (40 gallons)
Geographic origins: Red Sea, Western Indian Ocean.

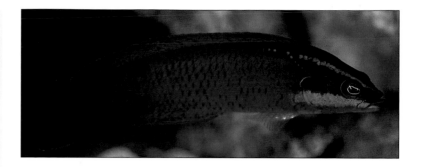

Other dottybacks

Here are two more varieties of *Pseudochromis* dottyback to consider.

NEON DOTTYBACK
Pseudochromis aldabraensis

This species is bright orange with bright neon blue stripes along the back, on the head, and on the edges of the anal and caudal fins. It is quite a territorial species, best kept with larger tankmates.

Maximum size: 4 inches
Minimum tank size: 48 × 18 × 20 inches (75 gallons)
Geographic origins: Persian Gulf, Gulf of Oman, Indian Ocean from Pakistan to Sri Lanka.

STRAWBERRY DOTTYBACK
Pseudochromis porphyreus

Superficially similar to *P. fridmani*, solid magenta in color, this species lacks the dark band through the eye, the tail is rounded, and is stockier. It is a much more territorial species than *P. fridmani*.

Maximum size: 2.3 inches
Minimum tank size: 24 × 12 × 16 inches (20 gallons)
Geographic origins: Western Pacific Ocean, Philippines to Samoa.

SPRINGER'S DOTTYBACK
Pseudochromis springeri

Maximum size: 2 inches
Minimum tank size: 24 × 12 × 16 inches (20 gallons)
Geographic origins: Red Sea, Western Indian Ocean.

DIADEM DOTTYBACK
Pseudochromis diadema

Maximum size: 2.3 inches
Minimum tank size: 24 × 12 × 16 inches (20 gallons)
Geographic origins: Western Central Pacific Ocean.

FALSE GRAMMA
Pseudochromis paccagnellae

Maximum size: 2.7 inches
Minimum tank size: 36 × 18 × 16 inches (40 gallons)

Geographic origins: Western Pacific Ocean; Indonesia to Vanuatu.

Clownfishes

Clownfishes are members of the damselfish family (Pomacentridae), found in the Red Sea, Indian Ocean, and Pacific Ocean, and are famed for their symbiotic association with large sea anemones. Fortunately, as sea anemones are very difficult to keep in the aquarium, clownfishes do not require this relationship in the reef tank. Instead, they will often adopt as a home another invertebrate (usually without harming it) or even an inanimate object.

Clowns make excellent aquarium fishes, subject to one big condition—they must be tank-bred individuals. Wild clownfishes are very delicate and prone to parasitic infections, but tank-bred specimens are extremely robust. The four species featured here are usually fairly easy to find in tank-bred form. There are 24 other species, but most are only available in wild-caught form, and some are very hard to obtain at all. In some species, captive breeding has made different color morphs available.

Clownfishes are easy to feed, accepting most dried and frozen foods. They are best kept in pairs. It is easy to form a pair by buying two juvenile fishes. Juveniles are all immature males; the dominant fish in a group or pair will become female, while the other(s) will remain male. Spawning in the aquarium is not uncommon and it is possible (but not easy) to raise the young. Clownfishes do not usually cause their tankmates any problems—at least when not breeding.

COMMON CLOWN
Amphiprion ocellaris

Maximum size: 2 inches
Minimum tank size:
24 × 12 × 12 inches (15 gallons)

Geographic origins: Indian Ocean and Western Pacific Ocean.

The amount of black coloration varies among percula clownfish.

PERCULA CLOWN
Amphiprion percula

Maximum size: 2 inches
Minimum tank size: 24 × 12 × 12 inches (15 gallons)

Geographic origins: Western Pacific Ocean.

CLARK'S CLOWN
Amphiprion clarkii

Maximum size: 6 inches
Minimum tank size:
48 × 18 × 20 inches (75 gallons)

Geographic origins: Indian Ocean and Western Pacific Ocean.

Left: *This is the similar, but slightly larger, orange skunk clown, Amphiprion sandaracinos.*

PINK SKUNK CLOWN
Amphiprion perideraion

Maximum size: 3 inches

Minimum tank size: 24 × 12 × 16 inches (20 gallons)

Geographic origins: Western Pacific Ocean.

Hawkfishes

Hawkfishes are predators of fish and crustaceans. They hunt by perching on rocks or corals, watching their surroundings intently, then darting out to catch passing prey. They lack swimbladders and so cannot hover in midwater without a great deal of effort. They behave in the aquarium much as they do in the wild, but they seem to be very intelligent and learn quickly to recognize their keepers.

Most hawkfishes are too large to be kept safely in the average reef aquarium, as they will eat smaller fishes and shrimps. However, the two small species discussed here are only a threat to very small fishes (such as neon gobies) and tiny shrimps. Otherwise, they are excellent reef tank inhabitants, being very hardy, not usually aggressive towards other species, and easy to feed on dried or frozen foods.

FLAME HAWKFISH
Neocirrhites armatus

Maximum size: 3.5 inches

Minimum tank size: 36 × 12 × 16 inches (40 gallons)

Geographic origins: Pacific Ocean.

LONGNOSE HAWKFISH
Oxycirrhites typus

Maximum size: 4.7 inches

Minimum tank size: 48 × 18 × 20 inches (75 gallons)

Geographic origins: Red Sea, Indian and Pacific Oceans.

Wrasses

The wrasses (Labridae) are one of the largest families of fishes, found throughout the world. Species range from the tiny (2-inch) possum wrasses (*Wetmorella* species) to the huge 7½-foot, 418-pound Napoleon wrasse (*Cheilinus undulatus*). Among their number are many wonderful, colorful fishes for the aquarium, including the 22 species featured here.

Fairy wrasses (*Cirrhilabrus* species)

The fairy wrasses include some of the most dazzling fishes for the reef tank. They are more than beautiful, however; they are near-perfect aquarium inhabitants. They are very hardy, relatively peaceful, and harmless to invertebrates. Most are bold fishes that swim high in the water column looking for plankton. Males and females often look very different (males are generally more colorful) and it is possible to keep a male with a group of females. Males of the same species should not be kept together, but males of different species usually tolerate each other. Like most planktivorous species, fairy wrasses should be fed at least twice daily; they will accept a wide range of dried and frozen foods. Their only weakness is a propensity to jump from open tanks, so you should keep the aquarium well covered.

GOLDFLASH FAIRY WRASSE
Cirrhilabrus aurantidorsalis

Maximum size: 3.5 inches

Minimum tank size: 36 × 18 × 18 inches (50 gallons)

Geographic origins: Western Central Pacific Ocean.

SCOTT'S FAIRY WRASSE
Cirrhilabrus scottorum

Maximum size: 5 inches

Minimum tank size: 48 × 18 × 20 inches (75 gallons)

Geographic origins: Pacific Ocean, from Great Barrier Reef to Pitcairn Island.

A hardy and very beautiful fish that is ideal for the reef aquarium.

EXQUISITE FAIRY WRASSE
Cirrhilabrus exquisitus

Maximum size: 4.7 inches

Minimum tank size: 48 × 18 × 20 inches (75 gallons)

Geographic origins: Indo-Pacific Ocean: East Africa to Tuamoto.

RUBY HEAD FAIRY WRASSE
Cirrhilabrus rubrisquamis

Maximum size: 3 inches
Minimum tank size: 36 × 18 × 16 inches
(40 gallons)
Geographic origins: Western
Indian Ocean.

KOI FAIRY WRASSE
Cirrhilabrus solorensis

Maximum size: 4 inches
Minimum tank size: 48 × 18 × 20 inches (75 gallons)
Geographic origins: Western Central Pacific Ocean.

Other wrasses

YELLOWSTREAK FAIRY WRASSE
Cirrhilabrus luteovittatus

Males of this beautiful species are bright purplish pink,
with a yellow streak along the flanks, a yellow dorsal
fin, and pelvic, anal, and caudal fins with purple rays.

Maximum size: 4.7 inches
Minimum tank size: 48 × 18 × 20 inches (75 gallons)
Geographic origins: Pacific Ocean: Pohnpei, Caroline
Islands, Marshall Islands.

BLUEHEAD FAIRY WRASSE
Cirrhilabrus cyanopleura

Males have a green head with a deep blue patch
immediately behind, reddish flanks, and a yellow
or white belly. The pelvic, anal, and caudal fins are
highlighted in purple, and dorsal fin is pink and yellow

Maximum size: 6 inches
Minimum tank size: 48 × 18 × 20 inches (75 gallons)
Geographic origins: Western Pacific Ocean.

SEA FIGHTER
Cirrhilabrus rubriventralis

Also known as the long-finned fairy wrasse.
Maximum size: 3 inches
Minimum tank size: 36 × 18 × 16 inches (40 gallons)
Geographic origins: Western Indian Ocean, Red Sea.

Halichoeres wrasses

Halichoeres is a large genus, with 75 known species that vary considerably in size, color, and suitability for the aquarium. *Halichoeres* wrasses are generally peaceful with unrelated species, and some of them (the banana wrasse *H. chrysus* and silver-belly wrasse *H. leucoxanthus*, for example) can be kept in groups. They are hardy and disease-resistant fishes. Smaller *Halichoeres* species are essentially harmless to aquarium invertebrates, but larger ones may eat small shrimps. However, they will eat worms (including bristleworms), small crustaceans, and small snails. Since these include a few "pest" species, such as red flatworms and the snails that parasitise tridacnid clams, these fishes can be very useful in the aquarium. They will also eat a wide range of frozen and dried foods. They are bold fish that spend their time cruising over the substrate looking for food. Their only specific requirement is a bed of fine sand, at least 2 inches deep, in which they can bury themselves to sleep or when frightened.

A group of banana wrasse make a colorful addition to the reef aquarium.

BANANA WRASSE
Halichoeres chrysus

Maximum size: 4.7 inches
Minimum tank size: 48 × 18 × 20 inches (75 gallons)
Geographic origins: Eastern Indian Ocean and Western Pacific Ocean.

Other wrasses

SILVER-BELLY WRASSE
Halichoeres leucoxanthus

This species is very similar to the banana wrasse, but instead of being solid yellow, it has a silvery white belly. Males develop a slight hump on the back of the head and facial markings similar to those of *H. iridis*.

Maximum size: 4.7 inches
Minimum tank size: 48 × 18 × 20 inches (75 gallons)
Geographic origins: Indian Ocean.

RADIANT WRASSE
Halichoeres iridis

The radiant wrasse does well in a peaceful reef aquarium.

Maximum size: 4.3 inches
Minimum tank size: 48 × 18 × 20 inches (75 gallons)
Geographic origins: Western Indian Ocean.

GREEN WRASSE
Halichoeres chloropterus

Maximum size: 7.5 inches
Minimum tank size: 72 × 18 × 22 inches (125 gallons)
Geographic origins: Indo-Malayan region of Western Central Pacific Ocean.

PINSTRIPE WRASSE
Halichoeres melanurus

Maximum size: 4.7 inches
Minimum tank size: 48 × 18 × 20 inches (75 gallons)
Geographic origins: Western Pacific Ocean, Japan to Tonga.

ORNATE WRASSE
Halichoeres ornatissimus

Maximum size: 7 inches
Minimum tank size: 48 × 24 × 24 inches (120 gallons)
Geographic origins: Eastern Indian Ocean, Pacific Ocean.

Hogfishes (*Bodianus* species)

Most hogfishes are hardy, medium to large, aggressive species that, although unlikely to harm corals, will eat shrimps, hermit crabs, snails, and possibly even small tridacnid clams. Fortunately, there are some smaller species and they are very attractive and unlikely to bother invertebrates. These small hogfishes are bold and spend most of their time in open water. They can become quite territorial when established in the aquarium, so should be added late in the stocking order. They will eat most foods. Keep their aquarium well covered, as they have a tendency to jump.

CANDY HOGFISH
Bodianus bimaculatus

Maximum size: 4 inches
Minimum tank size: 48 × 18 × 20 inches (75 gallons)
Geographic origins: Indo-Pacific Ocean.

REDSTRIPED HOGFISH
Bodianus species

Maximum size: 4 inches
Minimum tank size: 48 × 18 × 20 inches (75 gallons)
Geographic origins: West, Central, and South Pacific Ocean.

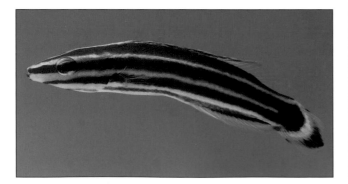

Flasher wrasses (*Paracheilinus* species)

Flasher wrasses are small, planktivorous species that hover in midwater. They are safe with all invertebrates. Males and females differ in coloration. Flasher wrasses can be kept in groups of one male and several females. Multiple males should only be kept in very large tanks. Flasher wrasses are peaceful fishes and need relatively quiet companions. They will eat most foods of appropriate size, but need feeding at least twice daily. Keep the tank covered, as they are jumpers.

LYRETAIL FLASHER WRASSE
Paracheilinus angulatus

Maximum size: 3 inches
Minimum tank size: 36 × 18 × 16 inches (40 gallons)
Geographic origins: Western Central Pacific Ocean.

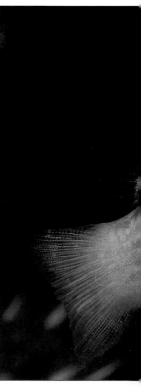

REDFIN FLASHER WRASSE
Paracheilinus carpenteri

Maximum size: 3 inches
Minimum tank size: 36 × 18 × 16 inches (40 gallons)
Geographic origins: Western Pacific Ocean.

Pseudocheilinus wrasses

Pseudocheilinus wrasses are rather secretive species that appreciate plenty of cover in the aquarium. They hunt tiny prey on the surface of rocks and sand. They will eat small parasitic snails that parasitize tridacnid clams. Most aquarium foods are readily accepted, too. They can be quite aggressive and should be kept singly, except in very large tanks.

MYSTERY WRASSE
Pseudocheilinus ocellatus

Maximum size: 4 inches
Minimum tank size: 36 × 18 × 16 inches (40 gallons)
Geographic origins: Western Central Pacific Ocean.

PAJAMA WRASSE
Pseudocheilinus hexataenia

Maximum size: 3 inches
Minimum tank size: 36 × 18 × 16 inches (40 gallons)

Geographic origins: Red Sea, Indian and Pacific Oceans.

The well-named pajama wrasse makes a colorful addition to the reef aquarium.

Other wrasses

BLUE FLASHER WRASSE
Paracheilinus cyaneus

The male of this species is unusually colored among flasher wrasses—the back and large dorsal fin are pale metallic blue, the flanks and belly purplish pink.

Maximum size: 2.9 inches
Minimum tank size: 36 × 18 × 16 inches (40 gallons)
Geographic origins: Western Central Pacific Ocean.

McCOSKER'S FLASHER WRASSE
Paracheilinus mccoskeri

Males of this species are yellow, with neon blue stripes along the flanks, a large bright red anal fin and a large yellow dorsal fin with neon blue markings. The head is reddish in some individuals.

Maximum size: 3 inches
Minimum tank size: 36 × 18 × 16 inches (40 gallons)
Geographic origins: Indian Ocean, Western Pacific Ocean.

Chapter Five

Setting up a reef aquarium

In this section we look in depth at creating a reef aquarium, beginning with the critical process of planning a system around its intended inhabitants, choosing the tank and equipment, and finding the right place to put it. In a detailed step-by-step sequence we set up a typical first reef aquarium, starting with an empty tank. Along the way, we demonstrate how to join pipes together, plumbing in a sump, filling the tank with purified water, adding marine salt, and measuring salinity. We set up chemical filtration, install a protein skimmer, and discuss how to choose and introduce live rock (a very important topic given the role of live rock at the heart of the aquarium's biological filtration system). We show how to build live rock displays that not only look attractive, but also create suitable niches for a variety of corals. We test the water to ensure the tank is ready to stock, then add a first selection of fishes and invertebrates.

Here you will also find advice on how to choose and buy healthy fishes and invertebrates, how to decide which ones and how many to keep, together with specific methods for adding different types of creatures to the aquarium—all of which have a tremendous impact on the success of the aquarium.

The work of setting up a reef aquarium is more than justified by the results.

Planning your aquarium

A successful reef aquarium does not happen by accident. The aquarium system needs to be planned carefully before you buy a tank or any of the equipment. At the very least, this can save a lot of expensive (and often difficult) revisions at a later stage. A single key principle underlies good aquarium system design: build the system around its intended inhabitants.

The fishes described in this book—a fraction of those that are commercially available in the aquarium trade—vary tremendously in size, from one to 16 inches when fully grown. Their diet ranges from algae to zooplankton and their preferred habitat may be open water or burrows in the sand. Activity levels vary too, from swimming continuously to spending most of the day perched on the substrate, and so on. The invertebrates covered here also vary greatly in their requirements for light and water movement; the types of substrates they grow on; their tolerance of suboptimal water quality; and their requirements for supplementary feeding, among other things.

Right: *These yellow watchman gobies (Cryptocentrus cinctus) live in burrows in the substrate, and the aquarium must have a sand bed to house them correctly.*

Some of the key aspects of the aquarium system that are affected by the choice of inhabitants are:

• Tank size: Large fishes need large tanks, and very active fishes need larger tanks than more sedentary species. Some invertebrates grow larger than others, too. Some corals can be pruned if they outgrow their allotted space, but

Below: *This large reef tank is well suited to its inhabitants: there is plenty of swimming space for the large fishes and powerful water movement has been provided for the branching stony corals.*

for others this is not usually possible. Tridacnid clams also vary tremendously in size, and *Tridacna derasa* as it grows will need a much larger aquarium than *T. crocea*. The tank you choose must be large enough to accommodate the creatures that will live in it.

• Water movement: Corals vary greatly in their requirements for water movement. Mushroom anemones, for example, generally prefer relatively gentle currents, whereas corals such as *Acropora* and *Porites* require powerful water flow. The tank must be fitted with circulation pumps to provide the appropriate degree of water movement for the corals to be stocked.

• Lighting: While lighting is not so critical for fishes and mobile invertebrates, sessile invertebrates vary tremendously in their light requirements. For example, *Acropora* corals generally require very intense lighting, whereas bubble corals (*Plerogyra* species) prefer less bright conditions. Most tridacnid clams are very difficult, if not impossible, to keep under fluorescent lighting—the highly colored *Tridacna crocea,* for example, will usually only do well under metal-halide lamps.

• Calcium: Fast-growing stony corals require abundant supplies of calcium and carbonates to build their skeletons. Reef aquariums housing large collections of such corals ideally need calcium reactors to keep up with their demands. Other methods of calcium and carbonate supplementation may suffice in tanks predominantly given over to soft corals, zoanthids, or mushroom anemones.

• Habitat: Some fishes, for example *Pseudochromis* dottybacks, *Pseudocheilinus* wrasses, and *Centropyge* angelfishes, live very close to hard substrates, and without sufficient rock in the aquarium will be shy and little seen. Others, such as *Cirrhilabrus* fairy wrasses, tangs, and *Genicanthus* angels, need plenty of open water for swimming. Others (many gobies, for example) need sand beds, perhaps of a certain type or depth, for building burrows. In each case,

the habitat required by the chosen fish needs to be provided in the aquarium.

These are just some of the aspects of the aquarium system design that are affected by the choice of inhabitants. Getting them right at the design stage is a great contributor to the long-term success of the aquarium.

Above: *An aquarium to house the* Tridacna crocea *clam requires very intense lighting, and a reliable method of keeping the calcium and alkalinity levels high.*

Below: *To feel secure, the pajama wrasse* (Pseudocheilinus hexataenia) *needs plenty of rocks and/or invertebrates as cover.*

Choosing an aquarium

Tanks are available in a huge range of sizes, shapes, and designs. Choosing the right one can have a significant impact on how well the aquarium works once it is set up.

How big?

A first decision is what size tank to use. While very large reef aquariums can look spectacular, they are very expensive and can be difficult to set up and run. At the opposite end of the scale, very small tanks—so-called nano-reefs—are inexpensive to buy and run, but can be tricky to keep.

A good size to aim for is in the range of 40 to 210 gallons. In terms of tank dimensions, this translates to somewhere between 36 × 18 × 16 inches (50 gallons) and 72 × 24 × 29 inches (210 gallons). Such tanks allow you to keep a wide range of fishes and invertebrates without being too unwieldy or expensive—although even within this range, total costs climb steeply as tank size increases.

What shape?

The shape of the tank, particularly the height and the front-to-back depth, can have an important effect on how well it works as a reef aquarium. Height is important because of the effects of water depth on light penetration (see page 179). It is easier to achieve intense lighting down to the base of a shallow tank than a tall one. However, tall tanks can look spectacular and provide good opportunities for invertebrates to grow vertically. It is possible to use a tall tank, either by incorporating more powerful lighting or by housing those invertebrates that require intense light in the upper parts of the tank.

Adequate front-to-back depth is important not only to provide good swimming space for fishes, but also because most corals grow outwards as well as upwards, so allow plenty of space in all three dimensions.

Aquariums are commercially available in a wide range of different shapes, not just the traditional rectangular tank. Cylinders, hexagons, pentagons designed to fit into room corners, bow-fronted and even wave-fronted tanks are all available. However, standard rectangular tanks and cube

tanks remain the best choice for reef aquariums. They offer a good combination of an undistorted view into the tank, with shapes that are easy to light well and provide with sufficient water movement.

Materials

Standard glass tanks are relatively inexpensive and easily available in a wide range of sizes and shapes; they are the choice of most reef aquarium keepers. The main alternative to standard glass is acrylic, which is much

POSSIBLE TANK SHAPES

There is a wide range of tank shapes on the market, and many can be used for reef aquariums, although some are better than others.

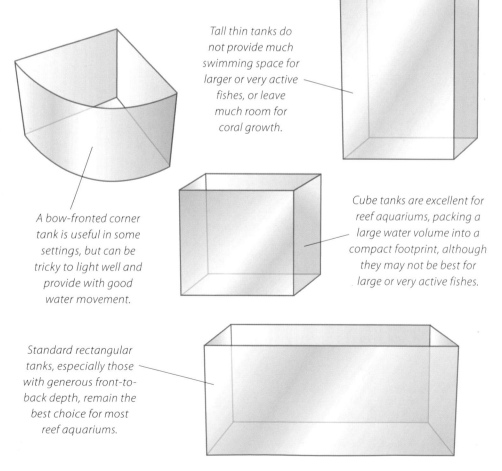

Tall thin tanks do not provide much swimming space for larger or very active fishes, or leave much room for coral growth.

A bow-fronted corner tank is useful in some settings, but can be tricky to light well and provide with good water movement.

Cube tanks are excellent for reef aquariums, packing a large water volume into a compact footprint, although they may not be best for large or very active fishes.

Standard rectangular tanks, especially those with generous front-to-back depth, remain the best choice for most reef aquariums.

FORESHORTENING IN AQUARIUMS

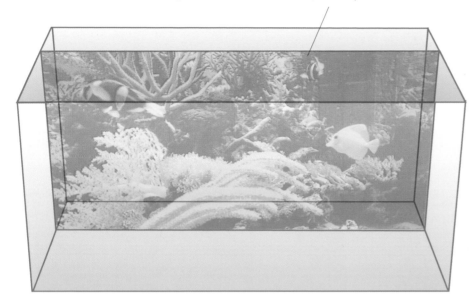

The foreshortening effect of water reduces the apparent breadth (front to back measurement) of an aquarium.

out of sight; increasing the system's total water volume; and potentially enhancing biological filtration and providing planktonic food by including a refugium (see pages 51–53). If it is not possible to fit a sump onto an aquarium, then using hang-on equipment, such as skimmers, power filters, and refugiums, can provide some of the benefits.

A sump can be a simple glass or plastic tank or a more elaborate construction with separate compartments within it. For example, there might be one for a skimmer, a second for a refugium, and another chamber for the return pump. A sump is usually positioned beneath the main tank, fed with water by gravity, with a pump to return it to the aquarium.

lighter, somewhat easier to drill for plumbing and visually clearer. However, acrylic is more expensive than glass and softer (so more easily scratched).

Although ready-made tanks are only available in acrylic and standard glass, a custom-built tank gives you the option of a front panel of optical quality glass. This has a low iron content and lacks the faint green tint of standard glass. It is more expensive than standard glass, but offers similar viewing quality to acrylic. However, the disadvantages of standard glass tanks should not be exaggerated—most fishkeepers find them perfectly adequate.

Sump or no sump

A sump gives you the opportunity to enhance the aquarium in a number of ways. These include housing equipment, such as heaters, skimmers, chemical filtration media, and even thermometers, thus keeping them

Thanks to the sump, the only technology visible in the aquarium is a circulation pump.

Within the cabinet, the sump holds the heater-thermostat, a skimmer, and refugium chamber.

Right: *This small reef aquarium has an attractive cabinet and top trim (which conceals a T5 fluorescent lighting unit). It uses an overflow weir (visible at the back) and an under-tank sump.*

FROM THE TANK TO THE SUMP AND BACK

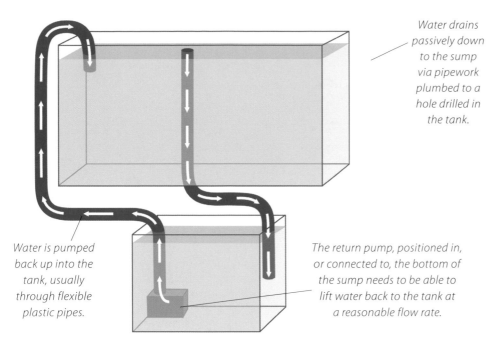

Water drains passively down to the sump via pipework plumbed to a hole drilled in the tank.

Water is pumped back up into the tank, usually through flexible plastic pipes.

The return pump, positioned in, or connected to, the bottom of the sump needs to be able to lift water back to the tank at a reasonable flow rate.

This prevents livestock being lost down the drain and determines the depth of water in the tank. Using a weir means that the water going to the sump is drawn from the surface, taking with it organic waste products, bacteria, and algae that tend to accumulate on the air-water interface. This continuous cleaning of the surface layer feeds skimmers or algae filters with water that is particularly rich in organic wastes, and improves gas exchange.

Return pumps

To return water from the sump to the display tank, you will need a reliable pump capable of lifting water the height from sump to tank (the "head"). It must also provide an adequate flow rate. A flow rate between tank and sump of 3–5 times the total tank volume per hour is generally recommended, but there is nothing wrong with a higher flow rate, provided the pipes delivering water to the sump can handle this.

Sump return pumps can be used to provide much of the tank's water movement. To maintain some flow

It is generally best if the sump is as large as possible, although if the sump is primarily intended to house equipment and keep it out of the main tank, it can be relatively small. Gravity-fed sumps need to have enough capacity (beyond the usual water level) to accept the water that will drain into the sump when the return pump stops.

Plumbing the sump

Under-tank sumps are fed with water by drains that run through holes drilled in the display tank. Glass tanks need to be drilled during construction. Holes can be drilled in the base or in the back or side panels of the tank.

Holes through the tank base are fitted with standpipes on bulkhead fittings, and the drains are out of sight under the tank. Holes through back or side panels use bulkhead fittings to connect them to exterior pipes. These require some space, so the tank cannot

be positioned right next to a wall, for example. The diameter of the holes (and the pipes that fit them) should be as large as possible. You can use flexible or rigid pipe to deliver water to the sump.

A watertight compartment is fitted around the hole or standpipe, with a weir, usually a glass or acrylic panel.

A SELECTION OF TANK SIZES, CAPACITIES AND WEIGHTS

Tank size (L × W × H)	Volume of water	Weight of water
36 × 18 × 16 inches	40 gallons	334 lbs
36 × 18 × 18 inches	50 gallons	418 lbs
36 × 18 × 24 inches	65 gallons	543 lbs
48 × 18 × 20 inches	75 gallons	626 lbs
48 × 18 × 24 inches	90 gallons	752 lbs
72 × 18 × 28 inches	150 gallons	1253 lbs
72 × 24 × 24 inches	180 gallons	1503 lbs

PREVENTING OVERFLOW

The sump needs enough capacity in reserve (on top of its usual water volume) to hold the water that will drain down when the return pump is switched off.

When the return pump is switched off, the water level in the tank will drain down through the return pipe until it reaches the outlet level.

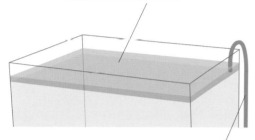

The sump needs to be large enough to hold this extra water volume.

Return pipe and outlet.

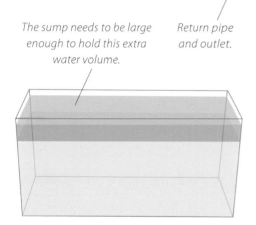

between tank and sump, even in the event of a pump failure, you can install more than one return pump.

Flexible plastic pipe is usually used to take water back to the aquarium, with either a simple rigid plastic outlet or a more complex unit that can be angled as required. Drill a small (.08 inch or so) hole in the outlet just below the water surface. This hole acts to break the siphon when the pump stops, limiting the volume of water that will drain into the sump. Purpose-made sump return outlets usually have siphon-break holes.

CALCULATING OVERFLOW VOLUME

Calculate the volume of water that will drain into the sump when the return pump stops as follows.

- Working in inches, multiply the length by the front-to-back width of the tank. This gives the tank surface area.

- Measure the depth from the water surface to the return pump outlet (or siphon-break hole if present). When the pump stops, the water will drain to this level.

- Multiply this depth by the surface area and divide this number by 231. This will give you the volume of the overflow in gallons.

- Multiply this by 1.5 (a safety factor) to get the volume the sump needs to hold.

Above: *This sump houses a skimmer, heater-thermostats, and chemical filtration media, and has plenty of space to accommodate the overflow volume.*

Siting the aquarium

To get the best from a reef aquarium, you must position it carefully and support it properly.

Choosing the spot

Locate the aquarium in a room that is large enough to allow the tank to be kept well clear of doors, windows, and radiators. Overheating is a particular issue for reef tanks, so avoiding heat sources (including sunlit windows) is very important. The room should ideally be a cool one, preferably (in the northern hemisphere, at least) north-facing. It is useful if there is a water supply nearby for filling the aquarium,

Below: This reef system is in a quiet corner, set back from the main route between two rooms and away from the direct sunlight coming through the window on the right.

performing partial water changes, and topping up evaporation losses, although in most houses there is little that can be done about the locations of kitchens and bathrooms.

The ideal site for a reef aquarium is quiet (away from sources of vibration and sudden shocks, such as banging doors) but not too quiet in terms of human traffic. Too noisy a location can result in jittery fishes, but so can too quiet a place—fishes that are exposed to little human activity can be nervous when people look at the tank. In contrast, most fishes easily become accustomed to the regular presence of people around them.

As well as the impact of the location on the aquarium, you need to consider the effects of the aquarium on the location. One key issue is the aquarium's

intense lighting. A television sharing a room with a reef tank will need careful positioning. Another consideration is the noise generated by the tank, most of which can be minimized by careful choice of technology. Skimmers are the main culprits with respect to noise. A third factor is humidity. This is only really a problem with open-topped aquariums; rooms housing such tanks need good ventilation. If you use cover glasses, the evaporation rate is much lower and increased air humidity is rarely a problem.

Supporting the aquarium

All aquariums are heavy—the water alone weighs over 8.35 pounds per gallon. To put this into perspective, even a 36 × 12 × 16-inch (30-gallon) tank will weigh as much as a heavy-weight boxer. A tank measuring

CHOOSING A LOCATION FOR YOUR AQUARIUM

Humidity can be a problem

The humidity in the room can be increased considerably by evaporation of water from open-topped aquariums. Using cover glasses can minimize this effect.

Avoid direct sunlight

The light from direct sunlight may be beneficial to corals, but sunshine tends to heat the aquarium water, so reef tanks are best kept away from windows.

Glare from tank lights

Reef aquariums produce a great deal of light, and this needs to be considered with respect to the tank's position in the room, to avoid glare on TV screens, for example.

Keep away from busy doors

Slamming doors disturb fishes, and if the aquarium is too close to a busy door there is always the risk that someone will bump into it.

Not near fires and radiators

Keep reef aquariums away from sources of heat, such as radiators and fireplaces, which could make the tank temperature unstable.

Keep the tank secure

Cabinets can prevent children and pets gaining access to under-tank equipment—which could be bad for both them and the aquarium.

100

Not close to the TV

Noise, vibration, and even flashing images from televisions and other electronic equipment can disturb fishes in the aquarium.

Provide adequate support

Aquariums are seriously heavy—be sure that the floor can support their considerable weight.

72 × 24 × 24 inches (180 gallons) with a sump weighs almost two thousand pounds. Aquarium cabinets are designed to take such loads and are available in a huge range of designs; other pieces of furniture may not be strong enough and should only be used with great caution. As well as supporting the tank, cabinets muffle any noise from equipment and prevent children and pets being able to access any under-tank technology.

The cabinet should be tall enough to accommodate any under-tank equipment (skimmers, for example), but many reef animals (fish and invertebrates) look best when viewed from slightly above, so very high cabinets are best avoided.

The floor beneath the cabinet must also be strong enough to take the considerable weight of the aquarium, with large tanks obviously presenting more of a challenge than smaller ones. Most concrete floors can support a large tank. So too can wooden floors with large joists, but ideally the tank should be on a supporting wall. On wooden floors,

it is best to position the tank so that its weight is supported on more than one joist. Consult a builder or structural engineer if you have any doubt as to the ability of a floor to support an aquarium, especially a large one.

Finally, the aquarium must be level; if it is not, this places uneven stress on the aquarium walls and joints, putting the tank at risk of breaking. If the floor is not level, use a stand or cabinet with adjustable feet or place slim pieces of wood under the base of the cabinet to establish the top level.

Preparing the tank

The glass tank used for the set up described here measures 45 inches wide, 24 inches front to back, and 30 inches high. It was supplied drilled through the base close to the rear wall, at the center, with an overflow weir around the hole. This weir gives a water depth of 28 inches. The tank has a gross capacity of approximately 127 gallons. While not excessively large, it offers a good water depth to allow upward growth of invertebrates, and a large base area to provide plenty of swimming space for the fishes. The tank is supported on a purpose-built cabinet that will hold a glass sump housing a skimmer, heater-thermostat, and chemical filtration media, together with a return pump.

Begin the setting up process by laying polystyrene tiles on top of the cabinet on which to place the tank, and on the base inside the cabinet on which to position the sump. To improve access for the installation of a sump, this cabinet has a removable central supporting strut at the front—it must be replaced before the tank is filled. Polystyrene tiles cushion the tank base and even out any slight irregularities in the top of the cabinet, ensuring that the tank is evenly supported. Unless specifically indicated by the tank and cabinet manufacturer, always use polystyrene tiles or similar material between the tank and the cabinet.

Position the tank on the top of the cabinet and place the sump carefully inside the base of the cabinet.

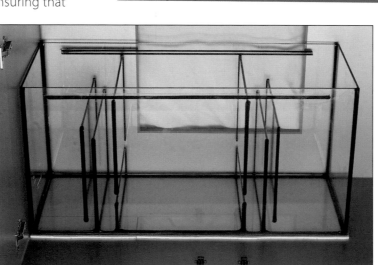

Above: The 127-gallon tank, with central overflow weir at the back, on a silver-finish cabinet with matching top trim.

Left: The sump in place in the cabinet. It has glass partitions separating different chambers that allow you to install an algae refugium in the central chamber. For this system it will be used as a simple sump to house aquarium equipment.

CUSHIONING THE TANK

Place polystyrene tiles on top of the cabinet to cushion the tank base. Stick them down with double-sided tape to prevent them moving when the tank is placed on top. Note the hole at the rear of the cabinet through which the pipe taking water to the sump will pass. Trim the tiles to allow for the aperture.

Above: Wipe down the inside surfaces of the tank with a damp cloth to remove any dust.

Below: Use a spirit level to check that the tank is level. It must be completely level, otherwise the tank walls may be subject to uneven stress, which could lead to failure of the joints. Making adjustments at a later date is virtually impossible.

A long spirit level is ideal for checking that the tank is level. It can easily be used to check both the front-to-back and left-to-right levels.

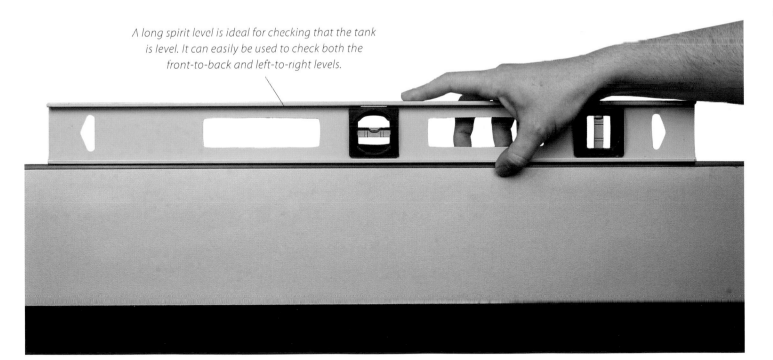

Although the tank has a hole in the base and an overflow weir, it still needs pipework to connect it to the sump. To do this, fit and connect a standpipe inside the weir, via bulkhead fittings, to a drain under the tank that will deliver the water to the sump. A pump is used to return water from the sump to the tank, via flexible plastic pipe connected to a rigid plastic outlet that delivers water close to the surface of the display tank. As water is pumped into the tank from the sump, it will overflow over the weir, run down the standpipe and back to the sump.

Create the connections between the tank and sump in two parts: a standpipe and a drain. The standpipe consists of a bulkhead fitting that goes through the hole in the tank base joined to a length of rigid plastic pipe with a connector. The drain is also made of rigid plastic pipe and constructed before the whole assembly is put together. Before fixing anything together permanently, do a dry run with both the standpipe and drain assemblies, i.e., without using any adhesives. This allows you to ensure that everything fits into place as it should.

When making joints using screw-threaded fittings, use a bead of silicon aquarium sealant on the threads. As you tighten the fittings, the sealant will be spread into a film over the threads and provide a waterproof connection. Similarly, where bulkhead fittings are tightened against the tank glass, make a seal between the fittings and the glass using a bead of silicon sealant. Secure push-fit connections in rigid plastic pipe with PVC pipe cement. When used properly, this will ensure a watertight joint. When using either silicon sealant or pipe cement, always follow the manufacturer's safety instructions.

CONNECTING THE SUMP AND THE TANK

The standpipe and drain assembly deliver water from the tank surface into the sump.

The standpipe sits in the center of the chamber behind the overflow weir.

A bulkhead fitting takes the pipework through the tank base—without any leaks!

The drain pipe delivers water into the sump.

The sump allows you to house equipment, such as heaters, skimmers, and chemical filtration, out of sight. It could also incorporate a refugium or algae filter.

ASSEMBLING THE STANDPIPE

Make up the standpipe before fitting it into the tank. You will need a threaded fitting, a flange nut, a connector, a length of rigid plastic pipe , silicon sealer, and PVC cement.

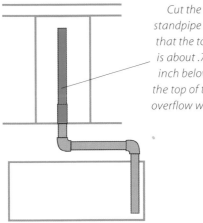

Cut the standpipe so that the top is about .75 inch below the top of the overflow weir.

1 This threaded fitting will become the base of the standpipe. Add silicon sealant before putting on a flange nut.

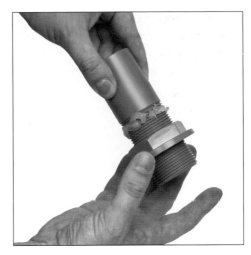

2 Tighten a flange nut up to the top of the thread, creating a bulkhead fitting that will seal the standpipe to the tank base.

3 Smear PVC cement over the unthreaded end of the fitting, in order to attach a push-fit connector.

4 Push the connector into place, twisting it slightly to smear the PVC cement over both parts of the joint.

5 Smear the open end of the connector with PVC cement, ready to glue in the main part of the standpipe.

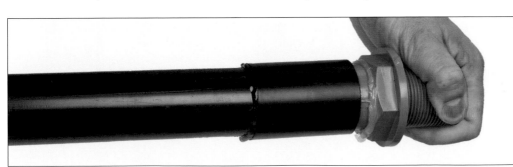

6 Cut a length of pipe to size and push it into the open end of the connector to complete the standpipe.

ASSEMBLING THE DRAINPIPE

The drainpipe takes water from the standpipe into the sump. On this tank it was constructed using rigid plastic pipe, but flexible pipe works equally well. When using flexible pipe, fit a hosetail onto the end of the standpipe, and attach the flexible pipe to it, using a hose clip for extra security.

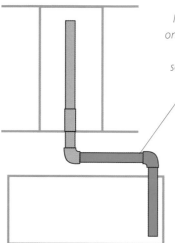

If the water is to enter at one end of the sump, create a bend in the drainpipe so that the outlet is in the correct place.

1 Coat the inside of one end of a pipe elbow with PVC cement.

2 Cut a length of rigid pipe to size. Insert it into the prepared side of the elbow.

3 Take a second length of rigid pipe and coat the end of it with PVC cement.

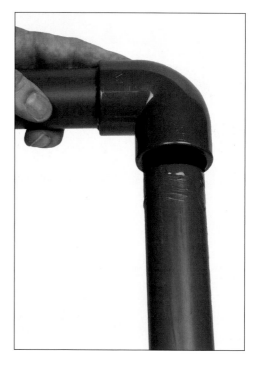

4 Push this end into the other side of the elbow to complete the drainpipe.

PUTTING IT ALL TOGETHER

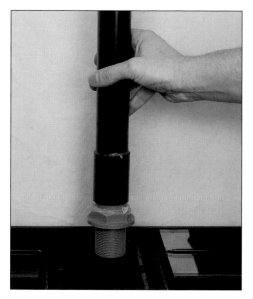

1 Lower the standpipe, with a bead of silicon sealer on the flange, into the weir chamber and through the hole in the tank base. Rotate it 180° to coat the silicon sealer between the flange and the glass of the tank base to ensure a good seal.

2 Fit another flange nut onto the thread at the end of the standpipe and tighten it up to the tank base. Place a bead of silicon sealer on the flange, with more silicon on the upper part of the screw thread. Tighten this nut to seal the standpipe in place.

3 Screw a threaded connector with an elbow already fitted (using silicon sealer on the threads) onto the thread at the base of the standpipe, again with more silicon sealer on the threads.

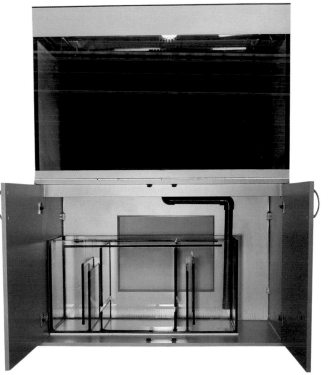

4 Secure the drainpipe into the elbow fitting, using PVC cement.

Right: The tank and sump with the drain in place, ready to feed water into the right-hand side of the sump. Leave the tank for one week to allow all the adhesives to set completely. On a cabinet of this design, replace the central supporting strut in the cabinet before filling the tank.

Filling the aquarium

Before filling the tank, make sure that it is sitting securely on its cabinet, that the sump (if used) is in position and that any plumbing is in place. Wipe the walls and base of the tank with a clean, damp cloth to remove any dust.

In most areas, tapwater is not suitable for the reef aquarium and needs to be purified (see pages 20–22). When filling the aquarium, you face the challenge of transporting the purified water from whatever system you are using to the tank. Often, it is a long way from the nearest water source. The least labor-intensive method is to run a pipe from the water purifier to the tank and fill it directly, but this is not always possible. The only alternative is to fill containers and carry them to the aquarium, emptying and refilling them as necessary.

Do not fill the aquarium completely; 90% full is a reasonable rule of thumb. This leaves some space to allow for the displacement caused by adding sand and live rock. It is more convenient to add slightly more water than necessary and then bale out any excess water after adding rock, sand, etc., than to underestimate the volume of water required. This would mean making up a separate batch of water to top up the aquarium to the required level.

When filling a large aquarium, bear in mind that water purifiers with limited capacity, such as deionizers and ion exchange systems, may become exhausted and need to be recharged (sometimes several times) before the tank is full. Periodically check the water these purifiers are producing to ensure that only adequately purified water goes into the aquarium—a simple dipstick test for nitrate is ideal.

Right: Fill the tank with purified water. Here, 6.5-gallon containers of RO water were used to transport water from a unit to the tank, but running a pipe directly into the tank, if possible, is less labor-intensive.

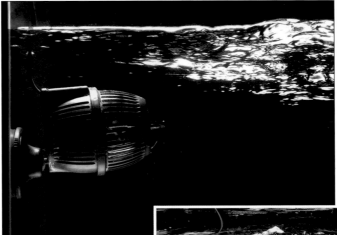

Left: To stir the salt and help it dissolve more rapidly, put one or more circulation pumps in the tank. Here, we used a 1,215-gallon per hour propeller pump for this purpose, but this is not its final location.

Right: Place a heater-thermostat on the bottom of the tank to heat the water to the required temperature. (It will eventually be moved to the sump.)

PUMPS

A large return pump (maximum flow rate 1,717 gallons per hour) delivers water from the sump to the tank. Data supplied with the pump suggests that with the approximate 3.2-foot head (distance between the water level in the sump and the outlet) the pump would deliver an actual flow rate of around 1,452 gallons per hour.

The intake strainer prevents debris being drawn in. Clean it regularly to maintain the pump's performance.

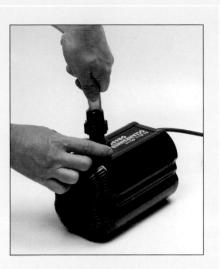

Right: *Fit flexible PVC pipe (here .62 inch) onto the pump outlet.*

RETURNING THE WATER TO THE TANK

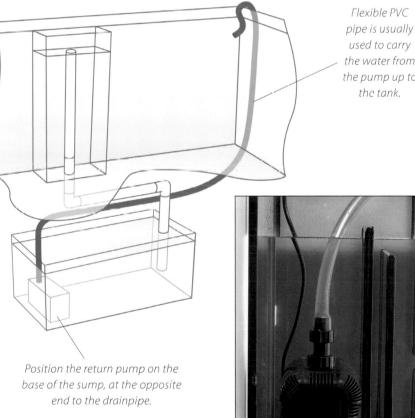

Flexible PVC pipe is usually used to carry the water from the pump up to the tank.

Position the return pump on the base of the sump, at the opposite end to the drainpipe.

Above: *The return pump and its associated pipework complete the water circulation loop between the tank and sump.*

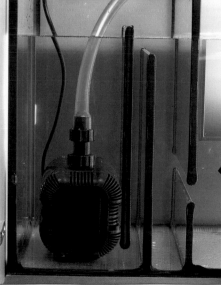

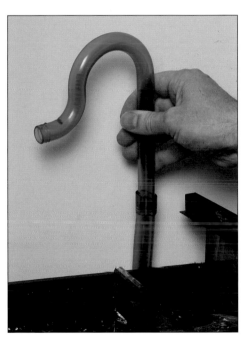

Above: *Fit a rigid PVC J-tube outlet to the end of the flexible pipe and hook it over a rear corner of the tank. Position it so that the flow of water is directed out into the center of the tank. The pump will provide much of the water movement required in the aquarium.*

Left: *The return pump in place in the left chamber of the sump.*

Adding the salt

Once the aquarium (and sump, if used) is almost full of water, you can add the marine salt. Water that is warm and moving vigorously will help the salt to mix and dissolve more rapidly. It is best, therefore, to put one or more heater-thermostats into the tank (even if these will eventually be positioned in the sump) and set them to the desired temperature for the tank. You can choose to turn on the heater-thermostats and allow the water to come to temperature before adding the salt or at the same time—this is not critical, although the salt will dissolve more quickly using the first method.

At this point you can place a circulation pump into the tank to mix the salt. Putting the pump on the bottom of the tank (even if this will not be its final position) or pointing its outlet at the tank base will help to prevent the salt from settling and should speed up the dissolving process. Start the return pump from the sump, so that water circulates between the tank and sump.

It is worth noting that some salt mixes contain a small quantity of insoluble material, so a little undissolved sediment is not a cause for concern. When all the salt has dissolved and the water is at approximately the right temperature, check the salinity using a hydrometer, refractometer, or electronic meter (see page 23). If the salinity is too low, add more salt. (The amount required can be calculated from the difference in the actual and desired salinities.) If it is too high, add more fresh water. When the salinity is at the correct level, the aquarium is ready to be brought to life by the addition of the live rock.

HOW MUCH SALT DO I NEED?

If you don't get a gallonage statement with your new tank, you can calculate the gallonage as follows: Multiply the length times the width times the height (depth). Dividing the answer by 231 will provide you with the gallon capacity of the tank.

- For example, with a 36 × 18 × 18-inch tank the multiplication comes to 11,664, and this divided by 231 gives you a 50 (.49) gallon figure. This is based on outside measurement. A 50-gallon tank actually holds a bit less water.

- A 50-gallon tank would roughly require 225 ounces (approximately 13 pounds) of salt to reach the required value. But, be sure and do your checking for specific gravity with your hydrometer or other equipment.

1 Weigh out the salt carefully, using a clean glass or food-grade plastic bowl; avoid metal bowls. Measure the large amounts needed for large tanks in batches.

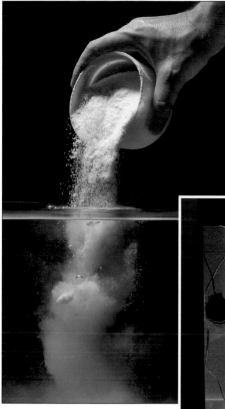

2 *Pour the salt into the water when the tank is almost full, with the water heated to approximately the correct temperature. This will reduce the time required for the salt to dissolve.*

HOW BUBBLES CAN AFFECT A READING

Bubbles adhering to the "needle" on a swing-needle hydrometer tend to push it up, giving a falsely high salinity and s.g. reading. A gentle tap will dislodge them.

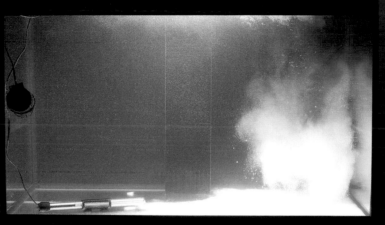

3 *Use a circulation pump (as here) or an airstone to stir the water. This will help to dissolve the salt. There may be some sediment left at the bottom of the tank even when the salt has fully dissolved.*

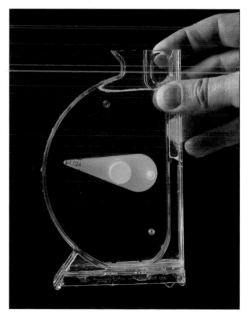

4 *A clean swing-needle hydrometer is a convenient way to check salinity and s.g. Rinse it in aquarium water before use.*

5 *Fill the hydrometer with water taken from the aquarium and tap it to dislodge any bubbles from the "needle."*

6 *Hold the hydrometer level and read off the specific gravity (s.g. shown on inner scale) and salinity (shown on outer scale).*

Chemical filtration and skimming

Incorporating a sump into the aquarium set-up enables two important parts of the aquarium's filtration system—chemical filtration and protein skimming—to operate inconspicuously, with no visual impact on the display tank. Chemical filtration, using either activated carbon or synthetic media, can be performed using so-called active or passive methods. In active methods, water is deliberately pumped through the media (in a canister filter for example), whereas in passive mode, water flows over the media but is not forced through it. Both methods can be used in a sump, the former by employing an internal power filter submerged in the sump to hold the media, the latter (as used in our example tank) by placing porous bags of media in the sump.

Many protein skimmers, including some large, powerful models are designed to be run either in, standing next to or hanging on, a sump. This is an ideal location for such a bulky piece of equipment, which could otherwise be very obtrusive. Running skimmers in or next to the sump is also ideal from the perspective of reducing noise. Skimmers are among the noisiest parts of the aquarium system, but most of their sound is muffled when they are housed in a cabinet beneath the tank. Of course, any skimmer that is going to be used in or on an under-tank sump needs to fit into the cabinet, so measure the available space and bear the dimensions in mind when choosing a skimmer.

ADDING CARBON

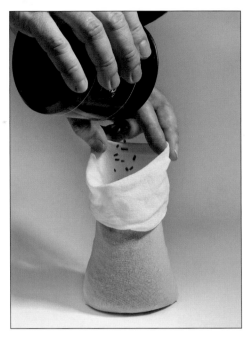

1 *Fill a media bag with carbon. In this system the carbon is used in "passive" mode, i.e., without forcing water through it. The feet of clean nylon tights can be used as home-made media bags.*

2 *When the bag is full, make a knot in the top to prevent carbon leaking into the sump. Some carbon may be spilled when filling media bags, so it is best to work in a shallow container or on a sheet of paper.*

3 *Once knotted, rinse the bag under fresh water to remove any carbon dust. Some bags have a drawstring closure, but it is best to make a knot for extra security.*

4 *Place the carbon bag in the sump, together with the heater-thermostat, which has been moved from the main tank to the central compartment of the sump.*

Left: The pump shows the design of the impeller, with multiple slim pegs rather than blades. These mix water and air together, breaking up the air into very fine bubbles that make for efficient skimming.

This skimmer has a large waste collection cup, so will need less frequent emptying than a smaller cup.

This pipe draws in water from the tank or sump, and air, feeding both to the pump for foam generation.

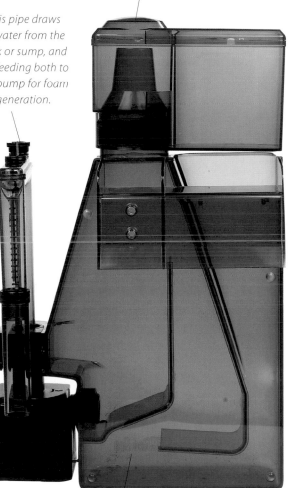

This needle-wheel skimmer is designed to operate either in hang-on or in-sump mode.

Above: The skimmer in action, set up to hang on the side of the sump and showing the dense bubbles within.

Adding live rock

Once the tank is filled with water at the right salinity and temperature, and circulation pumps and skimmer are running, the time has come to bring the tank to life by adding the live rock. The live rock is at the very heart of the reef aquarium and should be chosen carefully. If possible, visit several marine aquarium dealers and look at the rock they have on offer before making a decision as to which to use. It is best to be able to choose pieces of rock from a selection at a store. As well as making it possible to pick pieces that look right, it also means that you can check the rock to ensure that it is properly cured (using the "sniff test," for example). However, this might not always be possible; in some cases the dealer might have to order the rock, for example. (See also pages 36–43.)

Once the live rock is added to the aquarium, the exciting part of putting the aquarium together can begin.

Creating the bones of the underwater landscape by building live rock structures has a major impact on how the final aquarium will look, and it is worth spending plenty of time on this stage of the setting-up process in order to get it right. The ideal is to create loosely stacked (but still secure) rock structures that allow water to circulate easily and provide plenty of places to position corals and other sessile invertebrates.

When assembling live rock structures, it is a good idea not to build them too close to the rear wall of the tank. Leaving a space behind the rock structures allows better water circulation in the aquarium. It also creates secluded areas to which fish can retreat out of sight when they want to. Paradoxically, this means that they will probably spend more time out on display, if they are confident that there is secure cover nearby.

Above: Sniff each piece to detect any dead and decaying material. Properly cured rock smells, if at all, like fresh seafood—if there is any decaying material, the rock stinks.

Above: Unpacking Fiji live rock from its shipping box. Carefully inspect each piece to see what is growing on it and to look for any areas of possible concern.

LIGHTING UP THE AQUARIUM

When the live rock is in place, you can switch on the aquarium lighting for approximately 12 hours a day. This will help algae to grow, to the benefit of many small creatures living in the rock, and herbivorous fishes added later. On this system, intended to house corals with moderate lighting requirements, a luminaire with six T5 fluorescent tubes (three white and three actinic blue) was placed over the tank.

CREATING A "REEF" WITH LIVE ROCK

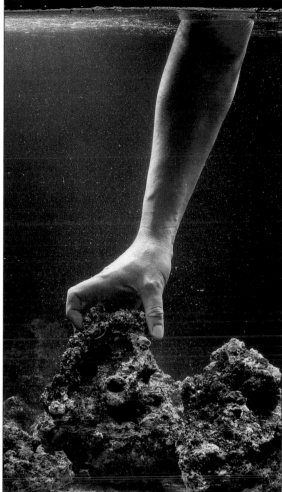

1 As you lower each piece of live rock into the tank, tumble it over a few times to release any trapped air bubbles that might otherwise kill organisms within the rock.

2 Do not attempt to build up the live rock at this stage. Simply lay each piece of rock gently onto the base of the tank. Do not drop pieces into the tank.

3 With all the live rock in the aquarium, spread out over the tank base, you can see how each piece looks underwater before starting to assemble the structure.

4 *Taking care to avoid rockfalls, pile up the pieces carefully to one side of the tank to clear space to start building. Build rock structures on the tank base before adding any sand. Use pieces of rock that will sit stably on the tank base as the foundation layer for structures.*

6 *With the foundations in place, you can begin building the next layer, trying pieces of rock until you find those that fit securely and look good together.*

7 *The rocks should be fitted together leaving plenty of space for water circulation through the structures.*

5 *Base rock is less attractive, but also cheaper, than Fiji live rock. It can be useful for filling an inconspicuous place at the rear of the structure.*

8 Assemble the structure carefully, ensuring that each piece of rock fits securely with the next. The irregular shapes of this Fiji rock make this straightforward, but in some cases you may need adhesives (two-part epoxy putty, for example) to secure the rocks.

9 Assembling a steep wall requires extra care; gentle slopes are easier to build, but the effect created here is more dramatic.

10 A total of about 88 pounds of live rock was used to create the completed structures. There is one large formation left of center, with a steep wall on the right-hand side, balanced by a small, free-standing structure on the right. The water is clouded by detritus from the rock.

11 Detritus from the rock has settled on the tank base. Siphon it out before adding any sand. The detritus is a mixture of organic material and small broken pieces of rock.

Adding sand

A sand bed brings numerous benefits to the reef aquarium (see pages 43–45), from improving biological filtration and providing habitat for a range of different creatures to making the tank look more attractive.

It is a good idea to add the sand or other bottom substrate after the live rock has been introduced and built up into the desired structures. If the live rock is sitting directly on the base of the tank, it means that the structures cannot be undermined by burrowing animals (which could happen if the sand bed was added first and the structures built on top of it).

You can use live sand or inert sand or a mixture of the two. Live sand is added straight from the pack, but inert sand needs to be washed carefully before use, which can be a time-consuming process, especially with finer sands. Live sands sometimes contain fine dusty material, and this may make the aquarium water cloudy at first, but this soon clears. For this set-up, we used inert and live sugar-fine aragonite sands, plus some coarser inert coral sand, to a final depth of approximately one inch.

HOW MUCH SAND DO I NEED?

The precise quantity of sand required depends on the particle size of the sand and the depth of bed planned for the reef aquarium. As a rough guide, a 1–1½-inch bed of sugar-fine sand (a good choice in most situations) requires around 77 pounds of sand per square yard of tank base.

PREPARING THE SAND

1 Inert (i.e., not live) sand, such as this coarse coral sand, must be washed before use. Place some in a bowl and add some water.

2 Swirl the sand around in the water, working your fingers through it, to stir up detritus and dust. Fragments of seaweed and plastic are often found in dry sand.

3 Pour off the water, refill the bowl and repeat the process. Thorough cleaning also removes any organic material that might rot in the aquarium.

4 The water will gradually clear as the sand gets cleaner—use as many changes of water as are required to get the sand clean.

5 When the sand is clean the water will be completely clear. Now it can be safely added to the aquarium. Repeat this cleaning process for further batches of sand until all the required material is clean.

Below: *Add sand using a plastic container, feeding it around the rock structures. Some sand will land on the rocks, but water currents will usually wash this away in time.*

Above: *Adding sand causes the water to become cloudy, due to the fine material in the live sand. This clears quickly as it settles out and is removed by the skimmer.*

Below: *Now the rock structures can be seen rising from the sand bed. Coarser coral sand was added on the left-hand side, where the strong current washed away the finer sand.*

Testing the water

With the live rock and the sand bed in place, the aquarium should be ready for stocking. However, before adding any fishes or invertebrates, it is wise to check that the tank is safe to stock by checking ammonia and nitrite levels. Adding cured live rock to the aquarium should be the equivalent of installing a fully functional biological filter, but if anything has died on or in the live rock (perhaps in transit between dealer and aquarium), large quantities of ammonia (and then nitrite) will be produced. This will mean that the aquarium cannot be stocked until levels have fallen back to zero, which will typically happen over a few days.

If you should detect any ammonia or nitrite at this stage, simply retest the aquarium every day or so until levels of both ammonia and nitrite have fallen to zero and have remained at zero for at least three days. The tank is then ready to stock.

When doing any water tests, follow the instructions carefully. To obtain accurate results, use the kits within their expiration date and store them as directed by the manufacturer. In particular, store color charts away from strong sunlight to prevent fading. Take great care that test reagents (some of which are quite toxic) do not find their way into the aquarium and keep test kits away from children and pets. It may be wise to wear protective gloves when performing some tests.

NITRITE TEST

1 *Always follow the instructions when carrying out water tests—not all products are the same. In this case, start by taking 5 ml of tank water and adding the required amount of test reagent—7 drops in this case.*

2 *Place the cap on the vial and shake it.*

3 *Now add the required number of drops of the second reagent to the vial.*

4 *Put the cap on the vial and shake it as before. In this test the color development begins immediately and is complete in 30–60 seconds.*

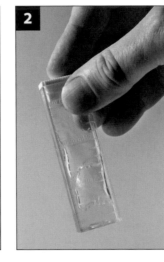

5 *Compare the vial against the color chart provided. It is clear that the nitrite concentration, although less than the 12.5 mg/liter that is the lowest value on the color chart, is above zero, so the tank is not yet ready to stock. In this test the vial is held against the color chart, but with some kits (where the color is not so strong) it is necessary to look down through the open vial from the top.*

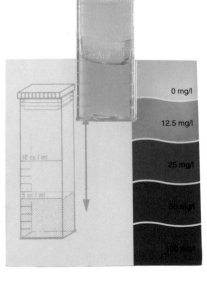

AMMONIA TEST

1 Rinse the vial in tank water to remove any tapwater left behind after washing.

2 Add the correct volume of water for the test (5 ml in this case).

3 Add the required number of drops of the first reagent as directed.

4 Place the cap on the vial and give it a shake.

5 Now add the required number of drops of the second reagent.

6 Place the cap on the vial and shake the vial again.

7 Add the required number of drops of the third test reagent to the vial.

8 Shake the vial again and (in this kit) leave it to stand for 20 minutes to allow the color to develop.

9 Reading the test against the color chart shows that the ammonia level in the sample is high, indicating that the tank is not yet ready to stock. This is probably the result of some undetected decaying matter somewhere in the live rock. Leave the tank unstocked and retest the water every day or so until both ammonia and nitrite levels are zero.

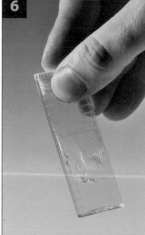

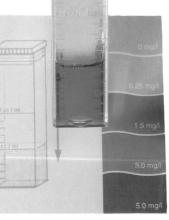

Preparing to stock the aquarium

With ammonia and nitrite levels consistently at zero, stocking the aquarium can begin. For most people, this is the most exciting part of setting up the aquarium, as it really brings the tank to life. In the reef aquarium, the stocking process is complicated by the fact that both fishes and invertebrates will be living in the tank. This adds an extra dimension to choosing, buying, and introducing the aquarium inhabitants into the tank, not to mention the expense involved. Stocking the aquarium is best done gradually, building up to the final population over several months, or even a year or more.

Choosing invertebrates

When selecting healthy invertebrates, the first step—and one of the most important—is to find a good marine dealer. Ideally, this should be close to home, making it easy to visit frequently and get a feel for the quality and selection of livestock. Personal recommendations are probably the best way to find such dealers, but information from fishkeeping magazines and Internet forums can also help. At a good dealer, most of the livestock should be of high quality.

When selecting corals, look for colonies that are firmly attached at the base to pieces of rock. Corals, whether soft corals, such as leather corals and their relatives, or stony corals, such as *Acropora, Montipora, Porites, Seriatopora,* and similar species, should have their polyps well expanded. Corals that consist of one or a few large polyps (for

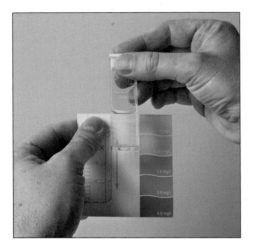

Left: After one week, both ammonia (as shown in the test here) and nitrite levels were consistently zero, so it was safe to begin stocking the aquarium.

Above: A variety of algae had started to grow on the live rock, including some green hair algae and brown macroalgae.

Above: The tank before the stocking process began, with the live rock structures and sand bed providing the basic underwater landscape, ready to receive the first invertebrates.

Right: Some tufts of green hair algae had also started to appear on the sand bed, along with some diatom growth.

example bubble corals, *Trachyphyllia, Euphyllia, Catalaphyllia, Caulastrea,* and similar species) should have their tissues and (if present) tentacles expanded. Stony corals should not have areas of exposed white skeleton. Zoanthids and star polyps should be fully open, with their tentacles extended. Mushroom anemones should have their discs spread out. There should be no signs of damage to the coral, or slimy white, brown, or gray patches, as these suggest infections. Try to avoid corals that have *Aiptasia* or "majano" anemones growing on them or on their substrate.

Indicators of health in other species of invertebrate vary between the different types of animal.

Echinoderms Avoid specimens with signs of physical damage. Faster-moving species (brittle stars and serpent stars, for example) should actively try to escape when being caught.

Crustaceans Mouthparts that are in constant, rapid movement are a good

Above: Healthy crustaceans, such as this peppermint shrimp (Lysmata wurdemanni), *should be lively, with busy mouthparts in constant motion.*

Below: Healthy snails should be moving over the substrate (or aquarium glass).

indicator of health, and livelier species, such as shrimps and crabs, should make determined efforts to avoid being caught. Hermit crabs should withdraw rapidly into their shells when disturbed. Crustaceans may have missing or damaged legs, claws, or other appendages. Although individuals in this condition are not ideal, in most cases the missing or damaged parts will regrow the next time the animal moults.

Snails Try to choose individuals that are moving around, or at least are fixed to the glass of their tank or to rocks.

Tridacnid clams Mantles should be well expanded, over the rims of their shells, and the animals should close their shells rapidly if disturbed, for example by shadows passing overhead. Individuals that have been resident in an aquarium for a long period may not do this, as they seem to become accustomed to such stimuli.

Buying and introducing corals

The corals are the most important inhabitants of the aquarium in terms of the way the tank looks. They need to be chosen and positioned with particular care, and not just for aesthetic reasons. Corals vary greatly in their requirements for light and water flow, and each one must be positioned appropriately if it is to thrive and grow. Find out as much as you can about the species you are thinking of stocking before buying. In the early stages of stocking the aquarium, most enthusiasts add corals a few at a time. Stocking the tank like this allows you to see how each coral settles into the aquarium and how it looks, before more are added, and lets you see how large each coral is when fully expanded. Another good practice is to stock lightly. Corals need plenty of space between them and their neighbors, and most coral colonies are capable of considerable, and often rapid, growth in the aquarium.

Adding corals to the aquarium

Marine animals cannot just be dumped out of their shipping bags into the aquarium. Many of them react badly to sudden changes in temperature, salinity, and other water conditions, and need to be acclimatized gently to their new surroundings.

A good method of acclimatizing new corals is illustrated on these pages. (Some other creatures need a special slow acclimatization procedure; see page 158.)

PREPARING INVERTEBRATES FOR TRANSPORT

Above: *Bought from a large aquarium dealer, a cabbage coral was first bagged underwater in the invertebrate sales tank (1). Plenty of water was included in the shipping bag (2), which was inflated with oxygen (3), then sealed (4). A second bag was added for security (5). Very spiky corals may be packed in a third or even fourth bag and sealed again (6). The bagged invert was then packed in a paper bag (7).*

ADDING INVERTEBRATES TO THE TANK

1 Float the shipping bag in the aquarium for 15–20 minutes. This will allow the temperature of the water in the bag to adjust to match that of the tank.

You can float several bags in the aquarium simultaneously, repeating steps 2 to 5 in rotation.

2 Open the bag. Corals may be growing on quite heavy pieces of rock, so take care that the bag does not sink. Add tank water to the water in the bag—a good quantity is approximately one-third of the volume of water in the bag. It can be difficult to estimate this, but slightly more or less is not harmful. Securely close the bag again and allow it to float for another five minutes.

Stages 3, 4, and 5 Repeat this process five times at about five-minute intervals, taking some of the water out of the bag, if necessary to avoid overfilling it. Leave the coral for another five minutes or so, then take it out of the bag and put it into the tank. The water from the bag can either be added to the aquarium or discarded; shipping water from an invertebrate will do no harm.

POSITIONING THE FINGER LEATHER CORAL

This finger leather coral, which is probably a Sinularia *species, was positioned about halfway up the rock structure, in good light and fairly strong currents. This area, where the currents from the return pump and the circulation pump meet, has turbulent water movement.*

Slow equalization for sensitive invertebrates

Some invertebrates, most notably crustaceans and echinoderms, are very sensitive to changes in salinity. They therefore require a slower and more careful acclimatization to the aquarium than corals or fishes.

The best method is to place the animal and some of its shipping water into a large container and then use a small-bore siphon (made of airline, for example) to add tank water drop by drop over perhaps two hours, until the volume of water in the container has increased 8- to 10-fold. Then transfer the animal to the aquarium. During this process, it is essential to keep the container warm, perhaps by standing or floating it in the tank or sump.

A simpler but still acceptable technique is to float the shipping bag in the tank for about 15 minutes to allow it to reach the same temperature as the tank. Then add tank water, in small quantities every five minutes over an hour or more, until the water in the bag has increased to perhaps three times the original volume. Remove some of the shipping water before starting to avoid overfilling the bag. Then transfer the animal to the tank.

Below: Some invertebrates, such as crustaceans and echinoderms (here a red serpent star), require a slow equilibration process, being sensitive to changes in salinity.

OTHER INVERTEBRATES THAT COULD BE ADDED

Although not shown here, a number of other invertebrates can, and should, be added to the aquarium. These other invertebrates are not just attractive and interesting, but also useful, in terms of controlling algae, clearing up uneaten food, and preventing detritus accumulation. Such creatures include small hermit crabs, algae-eating snails, and serpent stars and brittle stars. Various shrimps also make good aquarium inhabitants and can be added along with the corals and fishes.

Left: *The two button polyps (Zoanthus) colonies were placed high on the rock structure in strong water movement and bright light, and started to expand almost immediately after being added to the tank.*

Right: *Also in the upper part of the tank, the leather coral (Sarcophyton species) began to expand its polyps in response to the light and water flow.*

Left: *This green mushroom anemone colony is positioned quite low down, to the left of the rock structure. Here the water flow is relatively gentle, which suits these invertebrates.*

Left: *The finger leather coral (Sinularia species) is contracted, which is usual when soft corals are introduced to the aquarium. It will expand in a day or so.*

159

Introducing the fishes

Selecting healthy fishes can make a great contribution to the success of the reef aquarium. Finding good aquarium dealers is a great first step; they select their livestock with care, thus making life simpler for their customers. Choosing reliable aquarium species is also a sensible measure.

How many fishes?

It is commonly recommended that reef aquariums should only be stocked very lightly with fishes. However, modern systems for running reef aquariums, based on live rock and efficient skimmers, can support quite a heavy fish load. One good guideline is 2 inches fish length per 5 gallons of tank capacity. There are some caveats to this—it works better with more smaller fishes rather than fewer larger ones, and does not apply to what might be termed "gross feeders." These are predators, such as eels or groupers, that eat large meals and produce correspondingly large peaks of waste.

It is also a good idea to build up to this maximum stock level over a few months, perhaps leaving an interval of two weeks between additions of fishes. This allows the aquarium ecosystem (and not just the nitrifying and denitrifying bacteria) to adjust itself to the increase in biological load resulting from each new fish. Exceptions to this would be when pairs, groups, or shoals of fish need to be added. These should always be added together.

Fish size versus tank size

If an ultimate stocking level of 2 inches length of fish per 5 gallons of tank capacity is taken to its logical extreme, it would suggest that, for example, a 50-gallon tank could house a fish 20 inches long. As a typical 50-gallon tank might measure 36 inches long, 18 inches deep, and only 15 inches from front to back, this would clearly be absurd, as the 20-inch fish would not be able to turn around. It is clear that more guidance beyond the basic maximum stocking level is required in order to ensure reasonable swimming space for fishes. There are no absolute rules (fishes vary greatly in their activity levels), but it is possible to provide guidelines for "average" species (see panel on page 163).

Above: *Reef aquariums can support a generous population of fishes of various kinds, including (as here) algae grazers, planktivores, and small benthic predators.*

WHAT TO LOOK FOR WHEN BUYING FISH

In dealers' tanks, look for these indicators of good health.

- A fish that is alert, active, and interested in what is happening around it is usually healthy. However, the mix of species in a dealer's tank may affect the appearance of a fish. The presence of a more aggressive species may lead to a perfectly healthy fish being rather reclusive, but this should be fairly obvious in most cases.

- Good health in fishes is also indicated by body conformation—a healthy individual usually looks well fed, with a well-rounded belly and no "pinched" areas (particularly on the back, behind the head) that indicate muscle wasting caused by not feeding.

- Clear, bright eyes and intact fins, with no bloody streaks, cloudy areas, or frayed edges are another good indicator of health, although minor fin edge defects are not usually significant. The mouth should not be damaged, either—in some species, the mouthparts are so specialized that damage may doom the fish to starve.

- Healthy fishes have good, bright coloration, but it is important to note that many species can change the intensity of their colors according to lighting, background, and mood. In addition, the use of low levels of copper (for disease prevention) in dealers' tanks may also make fish less colorful. In some cases, the intensification of color when a fish is introduced to water without copper can be dramatic.

- In most species, apparently endless hunger is another good sign. It can be useful to ask to see a fish feeding before buying it.

- Finally (and most obviously) fishes should show no obvious signs of disease (skin lesions, rapid breathing, etc.).

PREPARING FISH FOR TRANSPORT

The yellow tang (Zebrasoma flavescens) was caught and bagged up ready to take home to the new aquarium. Having picked out a nice-looking individual, the fish was netted (1) and then bagged, with the bag inflated with oxygen (2). The bag was sealed with a rubber band (3,4). A second bag was added for security (5) and sealed in turn (6). The bagged fish was then placed inside an opaque paper bag (7), which provided some thermal insulation, but also meant that the fish was in the dark to induce it to sleep, making for a calmer journey home. Take your time over choosing fish and if you have any questions about the suitability of a species or condition of an individual specimen, do not hesitate to ask for advice. When buying subsequent fish, it is helpful to take with you a list of all the species currently in your aquarium.

ADDING THE YELLOW TANG TO THE AQUARIUM

Fishes are added to the aquarium in a similar way to corals, but take care not to add shipping water to the tank. The yellow tang was floated in the tank (1) for 20 minutes to allow temperatures to equalize. Then tank water was added to the water in the bag (2), which was then closed again and allowed to float for five minutes (3). This process was repeated five times and then the fish was netted in the bag (4) and transferred into the tank, swimming out of the net (5). The water in the bag was thrown away.

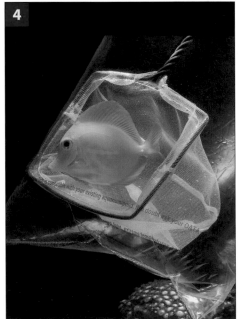

Fish stocking order

When keeping a selection of fishes, the order in which they are added to the aquarium can be important in maintaining a harmonious community, particularly when keeping species that are at all territorial. Knowing how territorial each fish is likely to be is a great help when stocking the tank.

Fishes that are already resident in an aquarium tend to feel that they have possession of the available territories. If a fish with a strong territorial instinct is already present, a meeker newcomer will find it very difficult to take over some of the territory. If, on the other hand, the meeker fish is already present when a more territorial fish is added, the meeker fish's lower aggression level is balanced to some degree by the advantage of possession of the territory. Also, fishes tend to accept other fishes that are already present in the aquarium when they are added, but to harass subsequent additions.

To use this situation to advantage, it is best to add the fishes in order from the least territorial to the most territorial. Of course, when doing this, it is important to bear in mind that not all fishes react territorially to all other fishes, so stocking a tang before a less aggressive species that is not a competitor of the tang (a clownfish, for example) should not be a problem, whereas adding two planktivorous species (which would be competitors) would need more careful management.

Pairs or groups of the same species (which should have more or less the same territoriality) should always be added simultaneously. This will help to prevent territorial conflicts that could arise if one fish was resident in the aquarium before others were added, and ensures that any group is always large enough to maintain harmony.

Quarantine

Choosing robust species that adapt well to aquarium life, as well as selecting strong specimens, can go a long way to ensuring that your reef aquarium's fishes live long and healthy lives.

For many fishes, potentially the most difficult part of their aquarium life is the period immediately after they are introduced to the tank. In the previous few days or weeks, wild-caught individuals will have been captured, transported perhaps half-way around the world or even further, held at an importer's facility and been transported again to a retail outlet. At this point, they may have been kept in water containing a low level of copper,

ADDING THE KOLE TANG TO THE AQUARIUM

This kole, or yellow-eyed, tang (Ctenochaetus strigosus) is one of two tangs in the tank. It has a different method of feeding to the yellow tang, browsing on detritus and ripping algal films from rocks with its bristle-like teeth. To minimize any territorial disputes between the tangs, they were added together.

TANK SIZE AND FISH SIZE

Tank size	Maximum fish length
36 × 18 × 16 inches or 36 × 18 × 18 inches or 48 × 18 × 24 inches or 48 × 13 × 16 inches	4 inches
48 × 18 × 20 inches	6 inches
48 × 18 × 20 inches or 48 × 24 × 24 inches	8 inches
72 × 18 × 22 inches	10 inches
72 × 24 × 24 inches or 72 × 24 × 29 inches	12 inches

ADDING THE WRASSES

Two very bold and colorful Cirrhilabrus *fairy wrasses—* C. rubriventralis *and* C. exquisitus *were also chosen for the featured tank. This is the sea fighter wrasse (*Cirrhilabrus rubriventralis*). By using a net to transfer the fish from the bag after equilibration, you avoid adding shipping water (which in this case contained copper from the dealer's tanks) to the display aquarium. Copper is harmful to invertebrates.*

ADDING THE GOBY

The tangerine-striped goby *(Amblyeleotris randalli)* was the first substrate-dwelling fish to be added to the aquarium. This is a peaceful species, unlikely to bother other gobies if they were to be added later.

ADDING THE CLOWNFISHES

*The two tank-bred juvenile common clowns (*Amphiprion ocellaris*) that were added to the aquarium should form a pair as they mature. These two juveniles were showing signs of forming into a pair within a few days. In a well-maintained aquarium they may spawn, although raising the young is very difficult.*

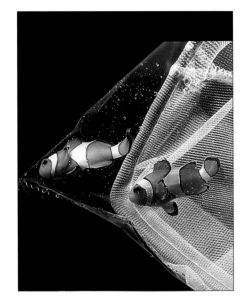

which, although it prevents parasitic infections, is mildly toxic. During some of this period, the fish will have fed little, to avoid its waste products polluting the shipping water. The fish then will go into a home aquarium, which will often have an existing population of fish that may act aggressively towards the newcomer. It is thus not hard to imagine why this is often the most difficult stage of a fish's aquarium life.

Implementing quarantine procedures can greatly ease this stage. Two weeks or so in a quarantine tank, with no competition from other fishes, allows the fish to recover from the rigors of transportation and to resume feeding properly. It also allows any diseases (which are more likely to occur at this stage than at any other time) to be identified and treated. The fish will then be fit, well-fed, and accustomed to aquarium life by the time it is added to the display tank.

A simple quarantine tank is inexpensive and easy to set up. Unless you are quarantining very large fishes or groups of several individuals of the same species, a tank measuring about 24 × 12 × 12 inches (15 gallons) will be big enough. Such tanks can often be accommodated in the cabinet under the main aquarium. Provide just a heater-thermostat, a light (a single fluorescent tube will do), and a thin layer of sand on the tank base. Add either a small quantity of living rock (which will provide both biological filtration and shelter), plus a powerhead for water circulation. Alternatively, install a small internal power filter, matured before the tank is stocked, and some inert decor.

Using the latter arrangement enables the tank to double as a hospital tank for treating fish with copper-based medications if necessary (see also page 195).

Some aquarium dealers routinely quarantine their fishes before offering them for sale, while others may be willing to accept a deposit to reserve a fish for later collection, thus providing what is in effect a quarantine service. Furthermore, fishes that have been resident in the dealer's tanks for a couple of weeks or more and are in good condition probably do not need any further quarantine.

Below: *The reef aquarium with its initial stock of fishes and invertebrates, already starting to look good. Some of the fish are exploring their new surroundings.*

The final display

At the end of the initial phase of stocking, the first corals were well-expanded, indicating that they were positioned correctly, and it was possible to see where further corals could be added. The fishes were healthy, with no evidence of any territorial conflict, even where this might be expected, namely between the fairy wrasses and between the tangs. The early growth of filamentous algae had been completely cleared by the two tangs.

The tank still has space for more fishes. Its 127-gallon capacity could ultimately hold around 47 inches total length of fish. The current residents, when grown to maximum adult size, would total around 29.5 inches total length, leaving up to 18 inches to be added. This would represent, for example, another four 4-inch fishes or six 3-inch fishes. More fairy wrasses of other species could be added or other hardy, peaceful planktivores, such as *Chromis*, firefishes or dartfishes, would be ideal.

This cave mouth is a calm area that might be a good place for a coral such as a Trachyphyllia brain coral or a bubble coral (Plerogyra sinuosa).

More mushroom anemones could be added in this area. Together with the existing colonies, they would create an attractive multicolored display.

This area, with strong currents and bright light, is a prime site to add more corals. Suitable choices would be more leather corals or branching soft corals, or zoanthids.

This rock could be left without invertebrates. Alternatively, low-growing, encrusting corals such as star polyps (Pachyclavularia) would look good.

Left: *The yellow tang, working in a team with the kole tang, will provide excellent control of nuisance algae, as well as being a beautiful, extrovert fish that is always on display.*

Below: *The* Cirrhilabrus exquisitus *fairy wrasse gets on well with the other fairy wrasse and swims out in open water, looking for plankton.*

Left: *The leather coral is fully expanded after a few weeks in the tank. It has the potential to grow into an impressive specimen; adding more of the same type of coral would look good.*

Below: *The mushroom anemones can be expected to reproduce and spread over the rocks and onto the sand bed beneath, forming a mixed, multicolored colony.*

Above: *Tank-bred common clowns* (Amphiprion ocellaris) *are always excellent and reliable reef aquarium fishes—hardy, peaceful, and safe with invertebrates.*

Chapter Six

Living with the reef

Creating a reef aquarium is just the beginning of what should be a long and fruitful relationship between the fishkeeper and his or her charges. Although modern methods of managing reef aquariums make keeping them much less labor intensive than in the past, some regular maintenance is required to keep the tank in top condition—in terms of both working well and looking good. In this section we discuss servicing the aquarium's equipment and using partial water changes to help maintain water quality, and look at some typical maintenance schedules.

As well as regular maintenance, the reef aquarium also needs some attention that no one would begrudge, namely feeding. Most fishkeepers would count feeding their tanks as a pleasure rather than a burden, and here we look at feeding both fishes and invertebrates, from choosing the best foods for different animals to deciding how and when to feed them.

In an ideal world, reef aquariums would be trouble-free, but although using modern techniques and choosing fish and invertebrates carefully can minimize the chances of problems, things do sometimes go wrong. Here we look at some of the more common reef aquarium problems—and provide some effective solutions. Finally, we take a look at how the reefs in the wild live with the aquarium hobby, exploring the impact of the aquarium trade on the wild reef environment.

Feeding the aquarium—every fishkeeper's favorite maintenance task.

In choosing foods for both fishes and invertebrates, the aim is to match their natural diet as closely as possible. While most prepared dry foods, such as flake and pellets, include a variety of ingredients designed to meet the needs of most species (although more specialized versions are available), other types of foods can be more specifically targeted.

Today, there is a wide range of marine aquarium foods available commercially. These divide into two main types, dry and frozen; most fishkeepers have to rely on these as staples, although some live foods can be used, too.

Dry foods

Dry foods include flakes and pellets of various types: there are small and large flakes, flakes designed for carnivores, others for herbivores, and so on. Pellets are also formulated in various sizes, with a range of ingredients to suit different types of fishes, and designed to float or sink. Dry foods of all kinds are highly nutritious, and with a little care in the choice of food (for example using flake with a high algae content for tangs) they make excellent staples for any fishes that accept them.

A specialized form of dry food that is extremely useful for feeding herbivorous fishes is dried seaweed, sold either in sheets or as flakes. This is an excellent staple food for tangs and other algae eaters. Sheets of dried algae can be put into clips attached to the tank glass and left for the fishes to graze. This type of food can also be used to feed algae-eating invertebrates.

Bear in mind that not all fishes will eat dry foods. In some species, visual recognition of food is evidently very important, so they will only accept food that resembles their natural prey.

Invertebrates such as shrimps, hermit crabs, and serpent stars also accept

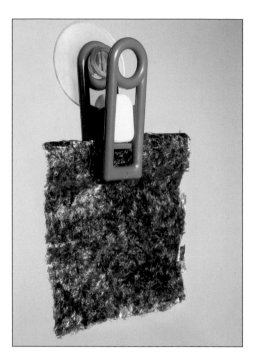

Above: *Dried seaweed sheets make an excellent staple diet for herbivores such as tangs. Using a feeding clip allows the fishes to graze on the seaweed at their leisure.*

both flake and pellets, and many corals will also eat these foods. Dry foods need to be fed fairly sparingly; their high nutritional content means that overfeeding the tank can lead to a considerable amount of waste being generated.

Frozen foods

A huge range of frozen foods is available. This includes many small crustaceans, such as red plankton (a type of small copepod), various forms of brine shrimp, mysis shrimp, and krill, which are ideal for feeding to planktivorous fishes, small predatory fishes, and to corals. Also available are various types of eggs (fish roe, lobster eggs, and oyster eggs, for example). Eggs of various species are a significant component of marine plankton, so

TYPES OF DRY FOOD

Small sinking pellets, ideal for smaller fishes.

Flake food is an excellent staple diet for most species.

Above: *Dry foods are very convenient to store and handle, and provide excellent nutrition for fishes.*

Larger pellets suit fish with bigger mouths.

Soft pellets can be pressed into lumps and stuck to the aquarium glass for browsers.

FROZEN FOODS

Brine shrimp
(Artemia)

Octopus

Krill

Microplankton

Above: *A pack of mixed frozen foods is ideal if you have only a few fish (so would only get through a limited quantity of each type of food) or limited freezer space.*

Right: *Blister packs are a very convenient way of packaging frozen foods, making "portion control" for fishes very easy. Thaw out frozen foods before adding them to the aquarium.*

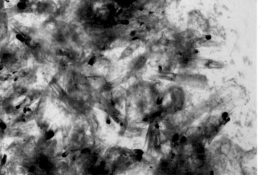

Left: *Frozen mysis shrimp is an excellent, highly nutritious food for marine fishes.*

This midwater feeder can capture food items as they fall through the water.

Below: *Frozen crustaceans are ideal for planktivores such as this firefish (Nemateleotris magnifica), providing a good representation of their natural diet.*

Above: *Brine shrimp (Artemia) are soft-bodied crustaceans relished by a wide range of fishes. The best types to use are those enriched with either Spirulina or omega-3 fatty acids. These are more nutritious than the basic product.*

these are very good for both fishes and corals. Although many other types are available, these are the most useful frozen foods for a reef aquarium housing the usual types of fishes and invertebrates.

Allow frozen foods to thaw before adding them to the aquarium, but for freshness it is best to use them as soon as they are thawed. Adding any juices produced during thawing may benefit corals and other invertebrates.

How to feed

How an aquarium needs to be fed depends on the requirements of its inhabitants. If there are planktivorous fishes, they should ideally be fed several small meals per day, although it may be possible to reduce this to two or three meals if the aquarium has a refugium (see page 53). This pattern of feeding also suits benthic predators and more flexible, opportunistic feeders will also do well with this type of regime.

Tangs and other algae grazers should ideally have access to food throughout the day. For these species, putting some dried algae in a clip in the morning and again in the evening is a good basic regime. The fishes will supplement this by grazing algae growing within the tank. Adding some algae-based flake once a day is also beneficial.

If the tank is stocked mainly with soft corals, it will probably need no supplementary feeding beyond providing food for the fishes. The juices of frozen foods contain plenty of fine particles, as well as dissolved proteins, and these seem to be enough for most soft corals. For stony corals, a light feed of frozen small crustaceans or roe, given after dark, perhaps once a week, is beneficial. A refugium can be very useful for feeding these corals.

RAIDING THE LARDER

There are plenty of foods intended for humans that can also contribute to the reef tank diet. Toasted seaweed sheets, nori, intended for wrapping sushi rolls, are a great staple food for tangs and other herbivores. Other types of dried seaweeds can be obtained from health food shops and oriental supermarkets, and salted fresh seaweed is occasionally available in delicatessens.

Left: *Dried seaweed sheets can be bought either packaged for aquarium use or as intended for human use, namely wrapping sushi rolls (when it is sold as nori).*

Below: *Dried seaweed sheets can be placed in clips so that fishes can graze on them at will. Tangs and other herbivorous fishes feed avidly on them.*

A SAMPLE FEEDING REGIME

This is a suggested daily feeding regime for a reef tank that holds a selection of soft and stony corals, some brittle stars, hermit crabs, and shrimps, plus tangs, planktivores, and benthic predators of various sizes.

- Morning: Dried algae sheet in feeding clip, frozen spirulina-enriched brine shrimp or frozen fish roe (alternate days).

- Lunchtime (or early evening if lunchtime is not possible): Frozen krill or frozen mysis (alternate days); more dried algae sheet in feeding clip.

- Late evening: Algae-based flake or carnivore flake (alternate days).

- Once a week: Frozen red plankton, lobster eggs, or fish roe, after lights out.

FRESH FOODS

A range of excellent fresh foods for reef aquarium fishes and invertebrates is available from a good fishmonger.

Chopped shrimp meat is relished by many fishes, and larger predators will eat whole prawns or shrimps.

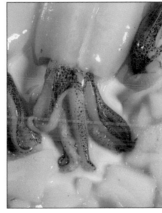

Above: This Lubbock's fairy wrasse (Cirrhilabrus lubbocki) is a primarily planktivorous species, although it will also hunt prey on the substrate.

Above: Squid is another useful and nutritious ingredient in a fresh seafood mix.

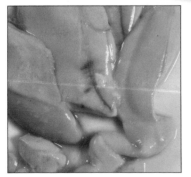

Left: Roe (shown here) and scallop meat can be chopped or blended, then fed to a variety of fishes and corals, either alone or mixed with other fresh ingredients.

FROM THE FISH MARKET

Fish markets can provide some good fresh foods for planktivores and small predators. Fresh (not smoked) hard cod roe, shrimps, prawn, and squid, chopped to an appropriate size for your fish, can all be used. When preparing fish for your own meals, you may find roe and these can also be used to feed the aquarium—if you can resist eating them yourself. With these foods, it is a good idea to freeze them overnight before use to kill any fish parasites that may be present.

Maintaining the aquarium

In order to keep a reef aquarium functioning well and looking good, with healthy, growing fishes and invertebrates, you will need to carry out a certain amount of maintenance. Fortunately, this is not too difficult or time consuming and need only take, on average, an hour or so each week.

Topping up

Water evaporates from the aquarium over time, with two important effects: the water level drops and, because only water—not salt—evaporates, the salinity of the water increases. Both can be dangerous. If the water level drops too far, heater-thermostats or pumps may be exposed to the air and break, or (if a sump is used) the cycling of water from tank to sump may be broken. If the salinity gets too high, it can be harmful

CLEANING OFF THE CALCIUM

One particular problem with reef aquariums is that the high calcium levels in the water tend to lead to equipment becoming encrusted with calcium carbonate, either as a result of precipitation or, more commonly, coralline algae growth. These deposits can be removed by soaking the affected equipment in a weak acid, such as spirit vinegar or kettle descaler. Even if this does not dissolve the encrustations completely, it will leave them easier to scrub off. After soaking, rinse the equipment very thoroughly before returning it to the aquarium.

MAINTAINING THE WATER LEVEL

Above: *The water level in the sump is too low; the return pump will start to suck in air and blow bubbles into the aquarium— topping up is required urgently. Automatic top-up systems (right) can prevent this.*

Right: *You can use a jug to add fresh RO water to replace evaporation losses in the aquarium.*

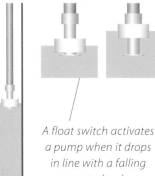

A float switch activates a pump when it drops in line with a falling water level.

An automatic top-up system, controlled by a float switch that turns on a pump to deliver fresh water when the level in the sump drops.

to both fishes and invertebrates.

To avoid problems caused by evaporation, fresh water must be added to the aquarium regularly. This can be done manually or by using a commercially available automatic top-up system. Aquariums vary greatly in their evaporation rates, with one determining factor being the use, or otherwise, of cover glasses. If cover glasses are used, the evaporation rate will be lower than if the tank is open-topped. For tanks with low evaporation rates, a once-weekly top-up works well. However, if you can top up every day, so much the better, as this will keep aquarium conditions very stable. Tanks with high evaporation rates will require

daily topping up or you can use an automatic system. More water tends to evaporate in hot weather than in cooler conditions.

Topping up should usually be done using tapwater, purified if necessary (see pages 20–22). You can work out how much water to add each time by marking the tank glass when it is at the correct level. Choose an inconspicuous spot somewhere on the back or side panel and use a marker pen or a strip of tape to make a mark at the water surface. This provides a level to aim for when topping up.

If you are using kalkwasser to maintain calcium levels (see page 56) it is normal to use this instead of water to replace evaporation losses. In this case, add the kalkwasser to the tank on a daily basis. Add it very slowly, using a dosing pump or very slow siphon. Do not pour it in all at once (which you can do if you are adding water).

Glass cleaning

One chore that most fishkeepers do very willingly is to clean algae off the aquarium glass. In reef aquariums this is made more difficult because the plaques of coralline algae that grow on the glass are much tougher than other algal films.

You can clean the glass using algae magnets (although these sometimes struggle with coralline algae) or scrapers. Other solutions include stainless steel scrubbing pads or old credit cards. These work well on glass tanks, but on acrylic tanks (which scratch easily) use only plastic scrapers. Keep scrapers well clear of silicon sealer at the corners of the tank.

Maintaining equipment

The equipment in the tank also needs regular maintenance—some items more than others. A typical schedule is described on page 178.

Skimmers

Regularly empty the skimmer's collection cup whenever it is full—once a week is usually enough. Thoroughly clean the foam tube in the center of the collection cup to remove the deposits of organic wastes that build up there. If left in place, these reduce the efficiency

Above: *A metal-bladed scraper is highly effective for cleaning algae, including coralline algae, from the aquarium glass, but keep it away from the silicon sealer at the corner joints.*

CLEANING THE SKIMMER

Skimmer collection cups need regular emptying and cleaning and not just to prevent overflows of unpleasant wastes. Check the level of protein waste (1). Empty the collection cup (2) and thoroughly clean both it (3) and the foam tube (4), using a cloth or sponge. Rinse well under running water (5). Keeping the foam tube clean ensures that skimming remains efficient, as deposits here tend to inhibit foam formation.

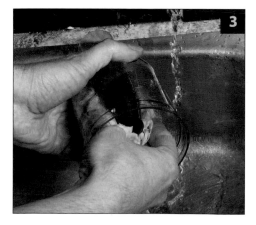

Left: Regular maintenance of calcium reactors will ensure a good supply of calcium and carbonates to meet the needs of stony corals, such as this Duncanopsammia axifuga, *in the reef aquarium.*

Wipe fluorescent tubes and the protective glass shields of metal-halide lights with a damp cloth occasionally to remove dust, salt deposits, and so on. Cleaning cover glasses is also important, as dust, salt deposits, and even algal growth can reduce the intensity of light reaching the aquarium. Plastic cover glasses can discolor after a few months under intense lighting. Replace them when they turn brown.

Both metal-halide and fluorescent lamps lose their intensity gradually, often undergoing a shift in output spectrum at the same time. Replace them regularly—once a year is usually effective.

New lamps are usually brighter (potentially quite a lot brighter) than old ones, so take care after replacing them, particularly when metal-halide lamps are involved. A sudden increase

of skimming. Weekly cleaning is a good routine. Every three months or so, strip down the skimmer completely and clean it thoroughly. The point at which air enters the venturi or pump intake needs particular attention, as salt crystals can build up in here, blocking the flow of air.

Pumps

Both circulation pumps and sump return pumps need regular cleaning. Whenever intake strainers begin to block (typically with algae or detritus), they should be cleared. Dismantle pumps and clean them thoroughly every three months.

Calcium reactors

Calcium reactors are among the most maintenance-intensive devices used on reef aquariums. Regularly check and adjust the water flow and rate of carbon dioxide injection. Every three months or so, strip down the reactor and give it a thorough clean. Replace the calcium

carbonate media when they are used up. Finally, refill or replace carbon dioxide cylinders when they are empty.

Lights

To get the most out of lighting of all kinds, it is important to keep lamps clean. Always turn off lamps and allow them to cool down before cleaning.

A TYPICAL MAINTENANCE SCHEDULE

Daily Feed and observe inhabitants; check equipment is working correctly; add kalkwasser (if used).

Weekly Empty and clean skimmer cup; test pH, alkalinity, calcium and nitrate; add calcium/buffer additives, if used; top up evaporation losses; clean lights and cover glasses; clean tank glass if necessary.

Monthly Clean circulation pumps; perform partial water changes.

Every three months Strip down and clean skimmers, calcium reactors, and pumps; replace chemical filtration media.

Annually Replace lamps.

in light intensity can be harmful to corals. So that the corals can adapt, it is a good idea to try to make the change more gradual by raising the light fitting after changing bulbs and gradually lowering it over a period of a couple of weeks. Alternatively, using a layer of some material to diffuse the light between the lamps and the water for the first week can be effective, as can shortening the photoperiod for the first few days and gradually returning it to normal. This procedure is not really necessary with fluorescents as their light is more diffuse, but when using a bank of tubes you can replace them one by one to minimize the impact of the change.

Chemical filtration media

Chemical filtration media such as carbon eventually become exhausted, and need to be replaced. Ideally, do not change all of the media at the same time (see page 54). Carbon is probably best replaced every three months or so. Change other chemical filter media on a similar schedule, with the exception of those that indicate (usually by a color change) when they are exhausted.

Water changes

One of the best ways to help maintain high water quality is to perform regular partial water changes. These dilute anything undesirable in the aquarium and help to replenish trace elements and other desirable substances. In most living rock-based aquariums, a reasonable routine is to change around 5% of the water per month.

To perform partial water changes, you must make up the required volume of artificial sea water to the same temperature and salinity as

MAXIMIZING YOUR LIGHTING

Light is emitted in all directions. Reflectors direct more of it into the tank.

Replace tubes once a year to maintain brightness levels. Change one at a time in multitube fittings.

Keep cover glasses clean to reduce light absorption. Replace plastic condensation trays when they become discolored.

Particles in the water "soak up" a large amount of light.

Some light is lost through the glass.

Changing metal-halide lamps will boost light intensity. Control the amount of illumination established corals and other invertebrates receive by diffusing the glare for a week or temporarily shortening the photoperiod.

the aquarium water. To do this, pour purified fresh water into a glass or food-grade plastic container. Add the correct quantity of salt and put a powerhead or an airstone into the container to mix the salt and water. In winter, add a small heater-thermostat set to the same temperature as the aquarium. Leave the water and salt to mix at least overnight, so that the salt dissolves completely and the water warms up to the correct temperature. Adjust the specific gravity if necessary by adding water or salt as required, so that it is as close as possible to that of the main tank. Siphon off the requisite amount of water from the aquarium (removing any obvious detritus in the process) and replace it with the new water.

REGULAR WATER TESTS

Below: Test pH regularly: this result shows that the pH level is in the correct range. This is a good indication that the aquarium system is working well.

Water testing

Testing some key water parameters on a regular basis allows you to detect problems before they become too severe and means that you can fine-tune the tank's maintenance routine to keep conditions at their best. Weekly testing of the following parameters is ideal.

• Specific gravity should be stable. Too high indicates that more topping-up is needed, but a decrease suggests a leak somewhere in the system or excessive topping up.

• The pH level should always be above 8.0, and when the lights are on at least 8.2, preferably higher. A lower level

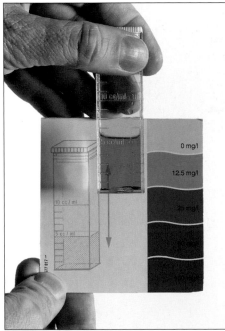

Above: Nitrate levels should be checked weekly. Here, the nitrate level is higher than it should be, so some remedial action is required to avoid problems.

suggests overstocking, overfeeding, or that you need to add a buffer.

• Alkalinity should be between 2.5 and 3.5 meq/liter (if you use an alkalinity kit) or 7–10 dKH (if you use a KH kit), preferably towards the top of this range. Falling alkalinity suggests the same problems as a low pH—alkalinity and pH are closely linked.

• Maintain calcium levels of at least 400 mg/liter; if levels drop too low, you may need to add larger doses of supplements, increase the size or frequency of water changes, or boost the calcium reactor's output.

• Keep nitrate levels as low as possible, although levels of up to 10mg/liter will probably not cause problems for most invertebrates. Rising nitrate levels suggest overstocking with fish, overfeeding, or that tap water purification is not working effectively.

Supplements and additives

There are many supplements and additives for reef aquariums on the market, but the value of most of them is unclear, particularly if you perform regular partial water changes. The exceptions are buffers and calcium supplements or two-part calcium and buffer combinations, if this is your preferred method of maintaining calcium and alkalinity levels. Adding these on a weekly basis is usually adequate. To determine the quantities required, be guided by testing calcium and alkalinity levels. If you are using other methods of maintaining calcium and alkalinity, such as running a calcium reactor, there should be no need for additives.

Left: This is a two-part supplement providing both calcium and a buffer to boost alkalinity, in solid form.

TESTING ALKALINITY

This alkalinity test uses a titration approach: the quantity of a reagent that needs to be added to the test sample to cause a color change indicates the alkalinity: the volume indicated on the syringe is compared with a chart provided with the kit to get the result.

Below: Dense populations of stony corals need plentiful supplies of calcium and carbonates in order to build their limestone (calcium carbonate) skeletons.

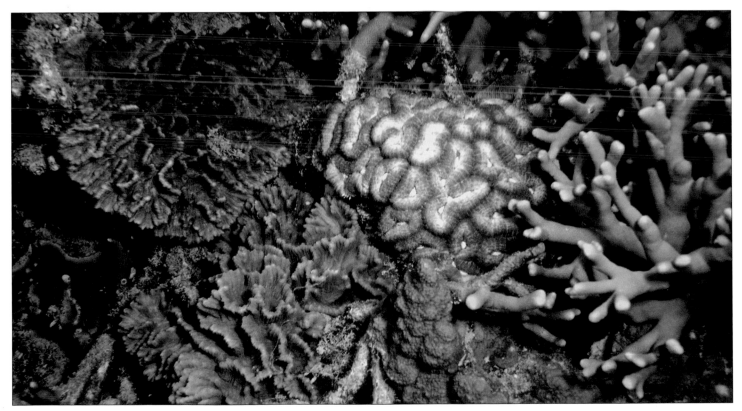

Solving problems

Most reef aquariums will have occasional problems, often related to undesirable creatures that arrive with live rock. Perhaps most common are nuisance anemones, which are like weeds in the reef aquarium garden.

Aiptasia

The most common reef aquarium pest is a small, semi-transparent, brown anemone known as *Aiptasia* that measures up to 1.2 inches across and 2 inches tall. *Aiptasia* can proliferate tremendously in a well-lit and well-fed aquarium and can damage corals growing close to them. They come into the aquarium on living rock and with corals. Even the tiniest piece of *Aiptasia* tissue can regenerate into a whole new anemone, so simply attempting to rip them off the substrate will not work and may even lead to an increase in their numbers. There are several

ways to tackle an *Aiptasia* problem, but it is important to act quickly, as small infestations are relatively easy to eliminate. Extensive *Aiptasia* problems can be beaten, but need much more effort and persistence.

Aiptasia can be eliminated biologically (using predators) and non-biologically (using mechanical and chemical techniques). Both small and large infestations can be tackled using biological methods, but non-biological techniques are only practical against small numbers of anemones.

Predators include both fishes and invertebrates. Plenty of fishes will eat *Aiptasia*, but the only more-or-less reef aquarium-safe species is the copperband butterfly (*Chelmon rostratus*), a difficult species to keep that can only be recommended for very experienced fishkeepers. The peppermint shrimp (*Lysmata wurdemanni*) is a much better bet for most fishkeepers. One or two

shrimps will be enough to deal with a few *Aiptasia*, particularly in smaller tanks. Larger populations of anemones need several shrimps—and patience. Peppermint shrimps are usually inexpensive and useful, harmless general scavengers. However, they are very sensitive to changes in salinity and need to be acclimatized to the aquarium very slowly. They are usually nocturnal; once added to the aquarium they might never be seen, the only evidence of their presence being the gradual disappearance of the *Aiptasia*. If progress is very slow, reduce feeding for a while; hungry shrimps are more inclined to eat *Aiptasia*.

Non-biological methods for dealing with *Aiptasia* need to be applied to individual anemones. One method uses epoxy putty. *Aiptasia* generally root into a small hole in the rock, into which they contract when disturbed. Carefully sealing over the hole with a blob of epoxy putty can prevent the anemone expanding and eventually kill it.

Other non-biological methods involve injecting *Aiptasia* with either lemon juice, a thick slurry of freshly mixed kalkwasser, or proprietary products. This is effective if performed correctly—successfully injected anemones often disintegrate immediately—but it is not easy, as the anemone must be injected in the central body cavity and it will usually try to contract into its hole when this is attempted.

Majano anemones

Tentatively identified as an *Anemonia* species, Majano anemones cause similar problems to *Aiptasia*, although they are less common. They resemble small mushroom anemones, typically brown

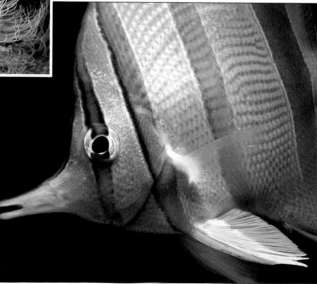

Left: Aiptasia *can reproduce rapidly, forming dense colonies that can damage neighboring invertebrates.*

Right: *The copperband butterflyfish* (Chelmon rostratus) *is the only fish predator of* Aiptasia *that is usually safe with corals. It is very attractive but not easy to keep.*

REMOVING A MAJANO ANEMONE

Majano anemones (below) are less common and proliferate less rapidly than Aiptasia, *but can still be a real nuisance, stinging and damaging nearby corals. A small specimen (right) appeared in the featured aquarium.*

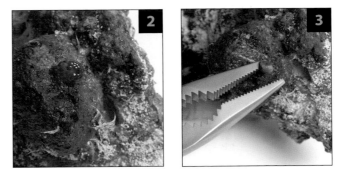

Having spotted the anemone on a piece of live rock (1), the rock was taken out of the tank—possible in this case, but not always (2), and the anemone removed using sharp-nosed pliers (3). The intention was to clip out a small area of rock with the anemone attached, but (as sometimes happens) the anemone detached itself from the rock and was discarded.

with lighter-colored tentacles. Majano anemones reproduce more slowly than *Aiptasia*, but in time they can build up large populations.

Majano anemones are harder to eliminate than *Aiptasia*, with no readily available, easy-to-keep predators that are safe with other invertebrates. Non-biological methods are therefore required. Injecting them with lemon juice, kalkwasser slurry, or commercial products can work, but is even trickier than for *Aiptasia*. Using epoxy putty to seal over them is not usually effective. Sometimes it is possible to chip out a small piece of the rock on which the anemone is growing and mechanically remove it this way.

Fortunately, Majano anemones do not appear to regenerate as easily as *Aiptasia*, so simple mechanical removal—pulling them off the rock—can sometimes work. Gripping the anemone with forceps will sometimes cause it to detach from the substrate

and it can be removed intact. Leaving a small piece of the anemone behind need not always matter. Sometimes it will grow back, but often it does not.

A range of other creatures that arrive accidentally on live rock can also cause problems in the reef aquarium.

Flatworms

Flatworms are common in a reef aquarium and they are usually unobtrusive and harmless. However,

certain types of flatworm can cause problems, notably the red flatworm *Convolutriloba retrogemma*. This species grows to about .2 inch long, and is a rusty red color, often with a prominent red spot. It can proliferate to plague proportions in a well-lit, well-fed aquarium and can cause problems by congregating on top of corals. It is possible that they feed on the corals directly, but any damage may only be due to irritation and shading—it is not clear which is the case. Some

Right: Red flatworms (Convolutriloba retrogemma) *can multiply into huge populations in well-lit, well-fed reef tanks. Fortunately, they have many fish predators, notably a variety of small wrasses.*

other flatworm species definitely eat or parasitize corals, and these too can proliferate in the aquarium.

Flatworm problems can usually be controlled using predators. The best are small wrasses that are safe to keep with other invertebrates. These include some attractive *Halichoeres* species, such as the banana wrasse (*H. chrysus*) and the pinstripe wrasse (*H. melanurus*), and it is likely that other *Halichoeres* species will also eat flatworms. Other flatworm predators include *Pseudocheilinus* wrasses, such as the pajama wrasse (*P. hexataenia*) and some fairy wrasses (*Cirrhilabrus* species).

Mantis shrimps

Mantis shrimps are fierce predators and some species grow to about 8 inches. They are potentially a threat to fishkeepers as well as their fish and invertebrates: they can inflict painful wounds with the specially modified claws they use to kill their prey. Fortunately, the larger species only appear in the aquarium very rarely. However, the smaller species occasionally find their way into the aquarium with live rock. These can sometimes be a threat to other aquarium inhabitants, but most small ones do not eat their tankmates, feeding instead on the food provided by the fishkeeper (which is easier prey, even for a super-predator such as a mantis shrimp). These smaller mantis shrimps are usually 1–2 inches long, and gray, brown, or green.

Mantis shrimps are most often spotted at feeding time, darting out to snatch food, or you may see their

Below: Fortunately, large mantis shrimps such as this are very rare accidental arrivals in the reef aquarium, but smaller species occasionally cause problems.

WHEN AND HOW TO REMOVE MANTIS SHRIMPS

It may prove necessary to remove mantis shrimps if the shrimp has harmed (or eaten) other tank inhabitants; is large or obviously growing fast; you are nervous about working in the aquarium with even a small mantis shrimp present. Catching mantis shrimps is difficult, as they are very intelligent (at least by crustacean standards) and quickly learn to avoid threats. Methods to try include

- Commercial traps.

- Predators, although most are not ideal to put into the reef tank and will eat other crustaceans. Candidates include bird wrasses (*Gomphosus varius* and *G. caeruleus*) and larger dottybacks (*Labracinus* and *Ogilbyina* species).

- Removing the rock that the shrimp lives in and extracting the shrimp from its hole by poking a skewer, flexible pipe, or strong wire into the hole and forcing the shrimp to come out (usually it emerges from another hole). Alternatively, put the rock into a bucket of aquarium water and bubble carbon dioxide into the water or pour carbonated mineral water into the hole. Either technique should encourage the shrimp to leave its lair.

When catching mantis shrimps it is best to protect your hands with strong gloves. Having caught the shrimp, you can then either kill it or, if you have a spare aquarium, keep it on its own, as these animals are very interesting to observe.

Above: *Seen from the front, this large mantis shrimp shows off the raptorial appendages that make it such a deadly predator—and a threat to aquarium keepers' fingers.*

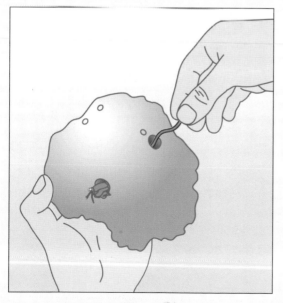

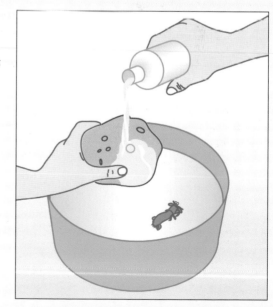

Left: *Poking a wire or skewer into a hole in live rock can sometimes persuade a mantis shrimp to leave by another exit.*

Right: *Pouring carbonated mineral water into a mantis shrimp's lair can often persuade it to leave, while doing minimal damage to other life in the rock.*

or external power filters packed with sponge, filter floss, or ceramic media to simple "socks" or baskets placed over the end of the pipe delivering water from tank to sump. Mechanical filters need frequent cleaning, otherwise the detritus will be trapped in the filter, with tank water being forced through it, and may begin to decompose, releasing nutrients into the water. Detritus that is kept suspended in the water column may also be eaten by filter-feeding animals, including corals.

Grazing to control algae

Both invertebrates and fishes can be employed to graze algae in the aquarium and these creatures are the key to controlling algae problems.

Grazing invertebrates

A variety of invertebrates graze algae and can be useful in preventing and controlling problems in the aquarium.

Snails There are many herbivorous marine snails that will not bother sessile invertebrates, but only a few species are commonly kept in the aquarium. Most snails prefer to graze on soft algae, with diatom films being particular favorites. Many snails also tackle short turfs of hair

Above: Small snails, such as this Lithopoma, *are useful grazers of algae, especially short turfs of hair algae and diatom films.*

algae, but long filaments seem to be beyond the abilities of most species.

The most popular algae-grazing snails for the aquarium are species of *Astraea*, *Lithopoma*, and *Turbo*. Confusingly, both *Astraea* and *Lithopoma* are frequently sold as "turbo snails." *Astraea* and *Lithopoma* are probably the most useful snails for most reef aquariums, thanks to their small size and hardiness. True *Turbo* snails are larger and not quite so hardy, but remain very useful grazers, as do the still larger *Trochus* and *Tectus* species. Larger snails need not only a greater supply of algae, but also sometimes cause problems by knocking over corals that are not securely fixed in place, or by breaking branches of more brittle species as they push past.

It is important not to overstock with snails, while at the same time adding enough of them to be effective in controlling algae. If you add too many, they will exhaust the available supply of algae and starve. Suggested stocking rates for a range of snails are one per 5

Above: Some small hermit crabs, such as this blue-knuckle hermit (Calcinus elegans), *are good algae grazers, as well as having other benefits for the reef aquarium.*

gallons of tank capacity for *Astraea* and *Lithopoma*, one per 13 gallons for *Turbo*, and one per 26–40 gallons for *Trochus* and *Tectus*.

Hermit crabs Various small hermit crabs are very useful grazers of nuisance algae, as well as being good general scavengers. Most species enthusiastically tackle short filaments of hair algae and diatom films and some will eat cyanobacteria. Hermits may also have beneficial effects on cyanobacterial blooms by stirring up the pockets of detritus on which the cyanobacteria tend to grow. Good species to stock for algae control include the scarlet hermit (*Paguristes cadenati*), the blue-legged hermit (*Clibanarius tricolor*), the blue-eyed hermit (*Calcinus laevimanus*), the blue knuckle hermit *(C. elegans)* and *Phimochirus holthuisi*. Sensible stocking levels for these species are around one crab per 4–13 gallons, according to the size of the crabs—you can stock more of the really tiny *Clibanarius tricolor* than of the much bigger *Calcinus laevimanus*.

Mithrax crabs, notably the emerald crab *(M. sculptus)*, the most commonly seen species, are also useful grazers of hair algae and bubble algae. They are usually safe with invertebrates, and are straightforward to keep.

Sea urchins are tremendously important algae grazers on coral reefs, but have only limited usefulness in the average reef aquarium. They have an unfortunate combination of being extremely strong and very clumsy, which means that they can knock

Above: Isopods are an important component of the reef aquarium microfauna.

Below: Copepods are important, if often unseen, algae grazers, particularly helping to control cyanobacteria.

Below: Emerald crabs (Mithrax sculptus) *are herbivorous crabs, among the few that can be kept in the reef aquarium. They are particularly useful as they eat bubble algae.*

over rockwork and dislodge corals. In addition, some species have sharp spines that can easily pierce the flesh of corals, as well as that of their keepers.

Microfauna is the term used to describe the small creatures that come into the aquarium with living rock, including a variety of crustaceans such as amphipods, copepods and isopods, as well as a range of worms, among other creatures. The microfauna make a major contribution to algae control

in the reef aquarium, particularly in keeping down cyanobacterial growth. They are often most active at night, and are the reason that in many tanks with cyanobacteria problems, the cyanobacterial mats seem almost to disappear overnight, but grow back during the day.

Grazing fishes

Several of the herbivorous fishes can make very useful algae grazers in the reef aquarium. It is important to note that the algae growing in the aquarium may not be enough to keep such fishes well fed, so supplementary feeding with algae-based foods will also be required. However, you need to strike a balance, so that the fishes are well nourished but remain sufficiently hungry that they will graze the algae growing in the aquarium. This is not usually difficult, as most herbivorous fishes need to eat plenty of algae and have an instinct to spend most of their time grazing.

Tangs and surgeonfishes Among the very best algae-grazing fish for the reef aquarium are the tangs and surgeons (see pages 95–101). However, not all tangs are good at controlling nuisance algae. Two genera are particularly useful. The most familiar of these are the *Zebrasoma* species: the yellow tang *(Z. flavescens)*, the purple or emperor tang *(Z. xanthurum)*, the two sailfin tangs *(Z. desjardinii* from the Red Sea and *Z. veliferum* from the Pacific)*, the less spectacular *Z. scopas* and the rarely seen, expensive black tang *(Z. rostratum)*. Most species grow to about 8 inches. (The sailfin tangs are larger, at up to 16 inches.) All will devour most hair algae and macroalgae and large individuals will often tackle bubble algae.

The other reliably useful group are the bristletooth tangs of the genus *Ctenochaetus*, for example the chevron tang *(C. hawaiiensis)*, the kole, or yellow-eyed, tang *(C. strigosus)*, the silver spot tang *(C. truncatus)*, and the orange-tipped bristletooth *(C. tominiensis)*. Most of the *Ctenochaetus* species are rather small for tangs, growing to 6–8 inches, the exception being the 11 inches chevron tang.

Zebrasoma and *Ctenochaetus* tangs feed on algae in different ways. *Zebrasoma* species snip off the ends of filamentous algae and macroalgae. *Ctenochaetus* species, in contrast, rasp algae away from hard substrates using large lips lined with small bristlelike teeth, preferring films of diatoms, cyanobacteria, and even dinoflagellates. They will eat filamentous algae, but prefer short turfs rather than the long strands that *Zebrasoma* species are happy to tackle.

Above: Zebrasoma *tangs, such as the purple tang* (Z. xanthurum) *are efficient grazers of filamentous algae, as can be seen from the very clean surface of the live rock, colonized only by coralline algae, in this aquarium.*

Left: Ctenochaetus *tangs, such as this* C. tominiensis, *scrape films of algae from rock and sand surfaces, a complementary grazing action to that of* Zebrasoma *species.*

Angelfishes Algae form an important part of the diet of many angelfishes. The suitability of angels for the reef aquarium varies; some species mainly graze algae and/or sponges, but others have broader appetites and will also eat, or at least nip at, tubeworms, and some corals. Angelfishes therefore need to be added to the reef aquarium with some caution. The *Centropyge* dwarf angels are potentially very useful reef aquarium algae grazers. They are much smaller than most tangs and can therefore be kept in smaller tanks. However, these angels are a little hit-and-miss in their utility as algae grazers, because most seem to prefer diatom films to hair algae, although some individuals also eat the latter very enthusiastically. The variation in taste for specific types of algae seems to lie more between individuals than between species—one coral beauty (*C. bispinosus*) may completely strip an aquarium of hair algae, whereas another will only graze diatom films. Not all *Centropyge* species are safe with corals or clams, and not all are easy to keep. Probably the best reef aquarium species is the coral beauty (*C. bispinosus*). The other *Centropyge* species featured in this book (see page 102) are also usually reliable reef tank inhabitants.

Blennies Some blennies are specialized algae grazers, and the diet of many other species includes a high proportion of algae. The relatively small size of many of these blennies makes them ideal as grazers for smaller tanks, where tangs could not be accommodated. Many blennies have wide mouths with broad lips covered in comblike teeth, which work in a similar way to those of *Ctenochaetus* tangs. Some good choices are the bicolor blenny (*Ecsenius bicolor*) and the Red Sea mimic blenny (*Ecsenius gravieri*). There are various other good algae-grazing *Ecsenius* species, but bear in mind that with most *Ecsenius* blennies (the prominent exception is the planktivorous *E. midas*) there is a small risk that they may try to scrape mucus from corals (especially fleshy stony corals). *Ecsenius* blennies are generally easy to keep and will accept many other foods as well as algae.

Among the more specialized algae-grazing blennies are the various *Salarias* species (including the well-named algae blenny *S. fasciatus*) and *Blenniella* and *Istiblennius* species, sometimes known as flymos or rockskippers. These are effective algae grazers, but because they are such specialized feeders they

Above: *In tanks that are too small for tangs, some blennies can make useful algae grazers. Most* Ecsenius *species, such as the bicolor blenny (*E. bicolor*), graze algae, but may also prey on fleshy stony corals.*

Right: *The algae blenny (*Salarias fasciatus*) is well-named. It has a voracious appetite for filamentous algae, and will need supplementary feeding once it has eaten all the algae in the aquarium.*

may starve if there is not sufficient hair algae present in the tank, unless they can be persuaded to eat a substitute, such as nori.

Maximizing the effectiveness of grazers—teamwork

Grazers vary both in their preferences for types of algae and in their feeding methods. Keeping a selection of grazers, both fish and invertebrates, can help to keep the aquarium as algae-free as possible, as their preferences and feeding methods are complementary. For example, most invertebrates cannot deal with long growths or thick tangles of hair algae, whereas grazing fish will eat these with relish, leaving short turfs that snails and hermit crabs will eat avidly. The combination of different tang species (which may need a large tank to avoid territorial squabbles) provides very effective algae grazing: a *Ctenochaetus* tang plus a *Zebrasoma* species together (because of their different ways of feeding) cover most nuisance algae.

Algae succession in the aquarium

Almost all new marine aquariums experience a series of algal blooms before they settle down to their final state and no one should worry about algae during the first few months of a new tank's life. In any new ecosystem, the first colonists to arrive are not usually destined to become the long-term occupants. Their growth changes the environment, allowing others to follow and probably to replace the pioneers. This process, known as succession, continues until the final plant community, known as the climax vegetation, becomes established.

In a new marine aquarium, spots of golden brown algae (diatom colonies) will appear on the rocks and sand within days of the tank being set up and may spread to cover the whole tank. The diatoms disappear a few days or weeks later, often to be replaced by films of red and green algae. These in turn are followed by green hair algae, which usually grows profusely at first, but eventually settles down to quite short turfs. If the calcium and alkalinity levels of the tank are as they should be, coralline algae will grow as

well, and together these will be the climax vegetation. This process is often modified somewhat in reef aquariums compared with more traditional marine tanks, as the more stringent control of algae nutrients and more intense grazing tends to favor coralline algae over hair algae, and to hasten the succession process.

It is worth noting that climax vegetation only remains stable in stable environmental conditions. Changes in conditions, be they increasing nutrient levels, decreases in light levels

A HELPING HAND

The fishkeeper can do a number of things to help grazing animals achieve the best effect in the aquarium:

- Snails and some grazing fishes quickly find out that it is easier to browse algae from glass or acrylic surfaces than from rocks. Keeping the tank glass clean forces grazers to work on other substrates.

- In the early stages of dealing with algae problems, it can help to place snails manually onto rocks with particularly abundant growth. The snails will often clean these rocks very quickly before moving on, and can be collected again and transferred to another "trouble spot."

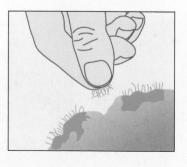

- Pulling out as much algae as possible by hand will give grazers a head start. All grazers, be they fish or invertebrates, find it easier to eat new growths of algae, rather than older, tougher material.

Above: *A copperband butterfly fish* (Chelmon rostratus) *showing signs of white spot, particularly in the tail fin.*

or changes in spectrum as bulbs age, decreased water flow, falling calcium levels, shifts in grazing patterns, or detritus accumulation, can all cause an established climax community of algae to change—often not for the better.

Fish diseases

In a well-managed reef aquarium, stocked with relatively hardy fishes (such as those described in this book), disease problems are likely to be very rare. There are two parasitic infections, namely white spot (*Cryptocaryon irritans infection*) and velvet disease (*Amyloodinium ocellatum*) that occasionally affect fishes in the reef aquarium. White spot, which is easier to treat, is much more common than velvet disease.

White spot infection shows as small white spots on the skin or fins of an affected fish. At first, just a few spots appear, but the numbers of spots increase as the infection progresses until the fish is densely covered in them. Eventually, the gills are affected and the fish may die. White spot is highly infectious (although some fishes seem to have a natural resistance), so if an infected fish is not treated promptly it is probable that its tankmates will soon be affected.

White spot can be treated reliably using two methods: low salinity and copper treatment. Both treatments should be performed in a separate treatment tank, as most invertebrates will not tolerate exposure to them.

Low salinity treatment is less toxic to fish than using copper. Variants of *Cryptocaryon irritans* that are tolerant of low salinity do exist, but these are rare and copper treatment can be used for infections that do not respond to low salinity.

There are many copper-based treatments for white spot on the market. A non-chelated copper treatment is the best type to use.

Velvet disease is a much more virulent infection than white spot and requires more rapid treatment. It is also more difficult to diagnose. The characteristic sprinkling of very fine spots on the skin, giving the "velvety" appearance described in textbooks, often occurs late in the infection. Earlier signs include rapid breathing (the gills are affected early in the course of the disease), scratching on rocks, lying inactive on the bottom of the tank (in a species that is usually active), and a loss of appetite (particularly if this is observed in combination with any of the other signs described here). Velvet disease should be treated with copper, following the manufacturer's instructions. Fortunately, this infection is rare in well-managed systems with sensibly chosen fishes.

Head and lateral line erosion (HLLE), also known as lateral line disease, is one further health problem that might be encountered. This is seen mainly in fishes that are adapted to eat a diet that includes a high proportion of algae, such as tangs and angelfishes. It manifests as loss of skin and (in severe cases) erosion of underlying tissue on the face and along the lateral line. It

LOW SALINITY TREATMENT PROCEDURE

Transfer the affected fish to the treatment tank (1). Decrease the specific gravity over several hours by removing saltwater and adding freshwater (2). On the first day, take the specific gravity down from the starting point (typically 1.025–1.027) to 1.017–1.018. Check the alkalinity and if necessary add a buffer to correct it. Continue the process on day two, going from s.g. 1.017–1.018 to 1.010–1.012. Check alkalinity again and adjust if needed. The white spot lesions should disappear quickly (typically in about 3 days). Keep the fish at this salinity for three weeks after the last lesions have gone (3). Check alkalinity regularly and correct as needed. Then adjust the specific gravity back to normal over two to three days by carrying out partial water changes with "extra-strength" seawater (water made up to s.g. of at least 1.030) (4). Observe the fish for one week to ensure the infection does not return (5). Return the fish to the display aquarium (6).

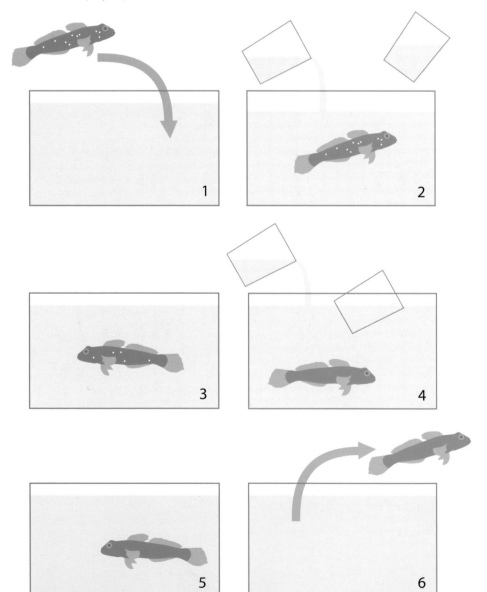

COPPER TREATMENT FOR FISH

Add a copper treatment (ideally a non-chelated formulation) to the water in the treatment tank to a level recommended by the manufacturer (1). Use a test kit to measure the copper level, as getting the concentration exactly right is very important. Transfer the affected fish to the treatment tank (2). Continue the treatment for about 10 days (this may vary with different products, so be sure to follow the instructions), checking the copper level regularly (3). When treatment is finished, remove the copper by water changes and/or filtration with carbon (4). Observe the fish for a week or so after treatment to be sure there is no recurrence (5), then return it to the main tank (6). Do not transfer any copper-contaminated water with the fish.

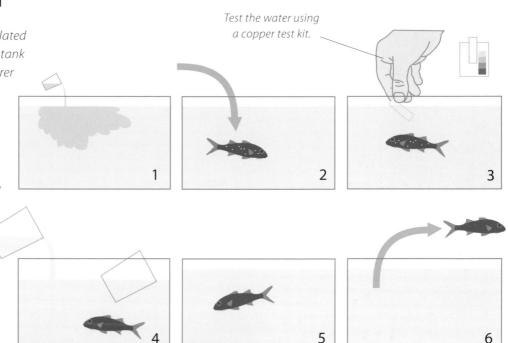

Test the water using a copper test kit.

Above: A brief period in quarantine before being added to the display tank can help a fish recover from the stresses of capture and importation, and allow any health problems to become evident in a setting where they are easier to treat.

A TREATMENT / QUARANTINE TANK

A treatment tank needs to be up and running as soon as an infection is observed. It is therefore a good idea to keep a small spare tank (24 × 12 × 16 inches [20 gallons] is adequate in most cases) running continuously. This can be used for quarantining new fishes, as well as treating sick ones. All that is needed is a simple aquarium with a heater-thermostat, an internal power filter providing biological filtration (which needs to be matured before the tank is used), a thin bed of sand, and some inert shelter, such as flowerpots or plastic pipe.

appears to be a nutritional deficiency, and can be prevented (and treated, at least in mild cases) by feeding a diet rich in algae.

Coral diseases

Just like fishes, corals are prone to a number of different diseases. In the reef aquarium, there are three conditions that are most likely to be observed.

Bleaching typically occurs after exposure of the coral to high temperatures, especially sudden temperature spikes. The corals expel their zooxanthellae and turn white. This is also a fairly common occurrence in the wild, in response to unusually high water temperatures. In the aquarium, faviid corals, especially *Platygyra* brain corals, seem to be particularly vulnerable to bleaching. Affected corals need no treatment: they usually recover after a few weeks or months as they capture new populations of zooxanthellae.

Rapid tissue necrosis (RTN) looks superficially like bleaching, but rather than just losing their zooxanthellae, affected corals lose their tissues, leaving behind bare white skeleton. This process can happen very rapidly, with the tissue disintegrating in a matter of hours. Typically, the tissue loss starts at the base of the coral and moves up the colony to the branch tips. Usually, all living tissue is lost from the colony, but sometimes the disease stops, leaving some areas intact.

The most commonly affected corals are *Acropora* species. Sometimes, several corals in the tank may be affected simultaneously. The cause of RTN is unknown, but it is possible that some type of infective agent is involved. A variety of bacteria have been found on affected corals, but it is unclear whether they are actually responsible for the problem. Colonies of *Acropora* that have been imported from the wild seem to be both more frequently and more severely affected than those grown in the aquarium from fragments. There is no reliably effective treatment for RTN, and it may be wise to remove affected corals from the aquarium to prevent possible transmission of the disease to other corals. It may be worth trying to take and grow on fragments from areas of affected corals that have not been reached by the necrosis—this way it may be possible to preserve the coral even though the original colony has been lost.

Brown jelly infections occur on a wide range of corals, often following trauma, such as physical damage, sudden changes in conditions, stinging by neighboring corals and so on. Areas of coral tissue disintegrate and are covered with a brown, mucoid material, hence the name. A protozoan called *Helicostoma* is often found in the "jelly," although it is not clear whether it is infecting the coral or simply eating dead tissue.

Brown jelly infections can be treated in many cases by using freshwater dips. Once corals have recovered, they will

A FRESHWATER DIP FOR CORALS

To prepare freshwater for dipping, fill a bowl or plastic bag with about 17 ounces of water. Add a small quantity of a marine buffer to bring the pH close to that of the tank (1). Float this in the aquarium for about 30 minutes so that it reaches the same temperature as the tank water (2). Dip either the whole affected coral or just the affected part into the fresh water and swirl it around gently. The coral can remain in the freshwater for up to five minutes (3). Then return the coral to the tank and discard the dipping water. In extreme cases, it may be necessary to carry out daily dips over a couple of weeks. Between dips, it may be useful to place the coral in an area with strong water movement. Dipping can be stopped when the coral tissue in the affected area develops a firm, clean edge and all brown jelly has gone.

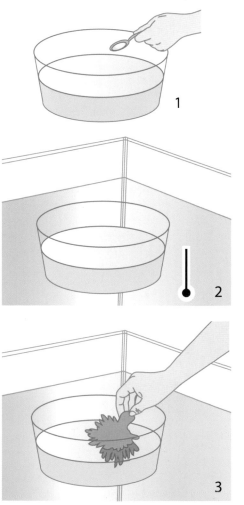

usually grow new tissue over any areas of bare skeleton. This infection (if it really is one) does not appear to spread easily between corals, but when moving or treating an affected specimen, it is best to avoid letting any of the gelatinous material come into contact with other corals.

Other coral problems

Sometimes corals do not always look as healthy or grow as well as they should, even though they do not have any specific diseases. Poor expansion of corals (with either the whole animal being rather shrunken or polyps failing to emerge) may be due to several causes

• The coral may not be positioned correctly with respect to either the light level or water flow. Move the coral, trying different places until you find one that suits it.

• Failure to expand only on one side suggests that the coral is being attacked by a neighboring coral; if there is physical damage on the affected side this is very likely to be the case. An alternative cause could be inadequate light on one side. The best solution in either case is to move the coral.

• If multiple corals fail to expand, this suggests that something is wrong with water conditions, especially if the corals were previously healthy. The best course of action is to check that tank parameters (pH, salinity, alkalinity, nitrate) are as they should be, and if necessary, correct any problems by making partial water changes. An alternative cause could be a sudden increase in light intensity, for example if metal-halide bulbs have just been changed. In this case, raise the lights or reduce the photoperiod temporarily (see pages 178–179).

• If a previously healthy coral fails to expand, it is possible that it is being pecked at by a fish. Some otherwise reef aquarium-safe fishes will sometimes selectively pick on a particular coral. Observe the tank carefully and if a fish is seen picking on the coral, be prepared to remove either the fish or the coral.

Below: *This* Acropora *colony is showing evidence of bleaching, probably the result of exposure to high water temperature.*

Conservation

Coral reefs worldwide are under threat from destructive fishing practices, mining for building materials, sedimentation resulting from deforestation, coastal development (particularly associated with tourism) and, potentially, increasing sea temperatures. Against this background, given that most of our aquarium inhabitants come from the wild, it is reasonable to question the impact of the aquarium hobby on coral reefs.

CITES regulations

Trade in stony corals and tridacnid clams is regulated under the Convention on International Trade in Endangered Species (CITES), which provides a mechanism for recording the numbers

Below: A healthy coral reef, with abundant vibrantly colored fishes and invertebrates— but under a variety of threats in many areas around the world.

of corals traded. No coral species are currently categorized as threatened with extinction, but both stony corals and tridacnid clams are listed under CITES Appendix II (species not necessarily threatened with extinction, but which could be vulnerable if they were over-exploited). With the exception of seahorses (which require specialized aquariums and are listed in CITES Appendix II) and the Bangaii cardinal, no marine aquarium fishes are considered to be at any risk of extinction.

Study of CITES records shows that, overall, the impact of collection of corals for the aquarium trade is very small in comparison with, for example, mining or fishing with explosives. Most corals kept in modern reef aquariums are species where harvest has little impact on the reef. These include leather corals, encrusting and branching soft corals, and fast-growing branching stony corals such as *Acropora, Seriatopora, Pocillopora, Montipora,* and so on. Of slightly more concern are solitary stony corals, such as *Trachyphyllia, Scolymia, Catalaphyllia,* and *Plerogyra,* where the whole coral is collected. Here, there is the potential for local over-collection, although their wide distribution in the wild would tend to preclude any major global impact on wild populations.

Marine aquaculture

Developments in marine aquaculture should reduce the already small impact of the aquarium trade on the reef environment. Throughout the tropics and subtropics, aquaculture facilities producing clams, corals, and live rock have become established. These also provide employment and revenue for people living in reef areas.

Collection of fishes and corals may even be beneficial to wild reefs by providing local people with an alternative source of income to fishing for food. When people are living in material poverty that few in affluent countries can begin to understand, it is impossible to blame them for doing whatever is necessary to scrape a living—even if that includes destructive fishing practices. Live fishes for the aquarium are worth much more than dead fishes for food and can provide a higher income to fishermen, and give them a stake in maintaining the reef in a healthy condition.

Away from the tropics, with no impact on wild reefs apart from the collection of mother colonies, aquarium coral propagation is also becoming important. Currently, apart

Above: *Tourism, even dive trips to explore reefs, can have deleterious environmental effects unless managed carefully.*

Below: *Lionfishes are now established on the East Coast of the USA, probably as a result of release from aquariums.*

from enthusiasts propagating corals to trade with each other, this mainly operates on a small commercial scale, but numerous species of corals are now available in tank-propagated form.

What can fishkeepers do?

Finally, there are two important things that aquarium keepers can do to ensure that the reef aquarium hobby remains environmentally sustainable. The first is to provide the best possible care for all reef tank inhabitants, so that they have long lives in the aquarium and their collection was not wasted.

The second is never to release any organism from the aquarium into the wild—this can have serious ecological consequences if the creatures in question survive. There are already examples of this: lionfishes, native to the Red Sea and Indian and Pacific Oceans, are established on the east coast of the United States, with an as yet unclear ecological impact. More seriously, *Caulerpa taxifolia*, a highly invasive type of macroalgae, has become established at sites in the Mediterranean and on the Southern Californian coast, with a major ecological impact. In both

THE CASE OF THE BANGGAI CARDINAL

The Banggai cardinal (Pterapogon kaudernii) is an excellent aquarium fish, but it has a very restricted geographical distribution, and fishing for the aquarium trade has had a significant impact on the wild population. Fortunately it is easy to breed in captivity; responsible fishkeepers should only purchase tank-bred individuals.

cases, deliberate or accidental releases from private or public aquariums seem to have been responsible for the introduction of these exotic species.

Right: *Although still mainly small-scale, captive propagation provides a supply of corals for the aquarium with no real impact on wild stocks.*

Lionfish
(Pterois volitans)

General Index

Credits

Unless otherwise stated, photographs have been taken by Geoff Rogers © Interpet Publishing.

The publishers would like to thank the following photographers for providing images, credited here by page number and position: (B) Bottom, (T) Top, (C) Centre, (BL) Bottom Left, etc.

Aqua Medic: 29(TR)

Bioquatic Photo—A.J. Nilsen, NO-4432 Hidrasund, Norway (email: bioquatic@biophoto.net. Website: www.biophoto.net): 7, 8, 10, 11, 12 (T, B), 14(T, B), 20, 24, 30(C), 31(TR), 33(BL), 38(TL, BL),

45(TR), 48(TL), 50, 56, 58(BR), 63(B), 64(TL, BL), 67(TL, C, BR), 69(BR), 70(BL), 71(T, B), 72(TL), 73 (T, B), 78(B), 79(T, B), 81(BL), 82(BC, BR), 83(TL, BC), 89(C), 90, 91(BL), 94, 95, 97(T), 100(C, B), 101(MiddleCR), 102(C), 103(BC, BR), 104(T, BL), 109(BL), 111(CR), 112(T), 113(T, C), 117(TR), 119(B), 120(CL), 124(TR), 125(T), 181(B), 182(CL), 183(BR), 184(CL, BR), 185(TR), 186(T, BL), 187(TR, BL), 189(TC, TR, BL), 197, 198–199(TC)

D-D Aquarium Solutions: 49(CR, BR)

Neil Hepworth: 80, 89(BR), 106(BR), 125(B), 130, 160

Phil Hunt: 23(T), 57(BR), 59(TL), 62, 66, 68(BL), 69(TR, CL, CR), 78(T), 81(BR), 87,

97(CR), 103(TR), 104(BR), 105(T), 107(C), 116(TR), 117(TL), 118(B), 129, 158(BL), 174(BR), 190(T)

Tristan Lougher: 53(TR), 74, 183(TL)

Scott Michael: 15, 16, 51(TR), 72(BR), 83(CR), 100(TL), 107(BL), 109(TR), 111(BL), 112(C), 116(C), 120(CR, BR), 198(BL)

Graeme Robinson: 17

Tropical Marine Centre: 57(TL)

Computer graphics by Stuart Watkinson © Interpet Publishing.

Publisher's acknowledgments

The publishers would like to thank Paul and Daryl at World of Fishes, Felbridge, East Grinstead, West Sussex, England (www.worldoffishes.com) for providing space, equipment, and livestock for the setting up sequences shown in the book. The aquarium featured was made by Seashell Aquariums Ltd, Chelmsford, Essex, England.

Thanks are also due to the following aquatic centers for hosting photography of livestock shown in the book: Cheshire Water Life, Northwich, Cheshire; Interfish, Ossett, West Yorkshire; Swallow Aquatics, East Harling, Norfolk; Swallow Aquatics, Gravesend, Kent; Swallow Aquatics, Rayleigh, Essex; Swallow Aquatics, Tenterden, Kent; Wharf Aquatics, Pinxton Nottinghamshire; Tropical Marine Centre, Chorleywood, Hertfordshire.

Publisher's note